Smart Logistics in Industry 5.0

Editor:
Hao Yu
Department of Industrial Engineering
UiT—The Arctic University of Norway, Norway

CRC Press
Taylor & Francis Group
Boca Raton London New York

CRC Press is an imprint of the
Taylor & Francis Group, an **informa** business
A SCIENCE PUBLISHERS BOOK

First edition published 2026
by CRC Press
2385 NW Executive Center Drive, Suite 320, Boca Raton FL 33431

and by CRC Press
4 Park Square, Milton Park, Abingdon, Oxon, OX14 4RN

CRC Press is an imprint of Taylor & Francis Group, LLC

Library of Congress Cataloging-in-Publication Data (applied for)

ISBN: 978-1-032-71798-2 (hbk)
ISBN: 978-1-032-73874-1 (pbk)
ISBN: 978-1-003-46645-1 (ebk)

DOI: 10.1201/9781003466451

Typeset in Times New Roman
by Prime Publishing Services

Preface

This book aims to discuss the impact of Industry 5.0 (the fifth Industrial Revolution) on the logistics sector. Today's rapid technological advances, e.g., internet of things (IoT), cyber physical system (CPS), artificial intelligence (AI), digital twin, etc., have radically transformed the industrial landscape and significantly influenced smart logistics operations. Diverting from a purely technology-centric focus in Industry 4.0, Industry 5.0 emphasizes primarily how these technological enablers can drive a harmonious societal change considering human-centricity, sustainability, and resilience. Thus, the interplay between technological enablers and the goals of Industry 5.0 in shaping the logistics sector, which is a traditionally labor-intensive industry, warrants thorough analysis.

With contributions from international authors, this book provides comprehensive discussions on essential Industry 5.0 enablers and their impact in smart logistics. First, this book explores distribution and logistics network design models and methods enabled by generative AI (GenAI) and digital twin technologies, data-driven smart transportation and logistics powered by machine learning and artificial neural networks, and the integration of social values and circular economy principles. Furthermore, it addresses extended reality in smart warehousing and highlights best practices for smart reverse logistics in a rapidly evolving industrial landscape of Industry 5.0. By introducing the latest theoretical developments, methodological advancements, and best practices encompassing both qualitative and quantitative methods, the authors present insights that move beyond current practices to highlight the future opportunities and challenges Industry 5.0 brings for disruptive change in smart logistics.

Reflecting cross-disciplinary approaches and broad topics related to smart logistics and Industry 5.0, this book will be of interest to a wide range of academic researchers and graduate students in engineering, supply chain and logistics, business administration, management, data science, and computer science. Moreover, it is also relevant to industrial professionals who seek to expand their knowledge of how new technologies can enhance smart and sustainable logistics practices.

Hao Yu is an Associate Professor and Program Leader of MSc in Industrial Engineering at the Faculty of Engineering Science and Technology at UiT—The Arctic University of Norway, Narvik, Norway. His research interests are real-problem-driven model development to solve complex decision-making problems across diverse fields. Currently, his work centers on integrating predictive, prescriptive, and descriptive analytics within a Smart Digital Logistics Twin in Industry 5.0.

Contents

Preface	*iii*
List of Contributors	*vii*
1. Distribution and Logistics Network Design in Industry 5.0: Models, Methods, and the Impact of Generative AI and Large Language Models	**1**
2. Distribution and Logistics Network Design Using Digital Twin: Concepts, Methods, and Applications	**31**
3. Artificial Intelligence and Data-driven Smart Transportation and Logistics: Unsupervised Learning Models and Ensemble Learning Models	**57**
4. Artificial Neural Network Methods in Smart Logistics: Concept, Methods, and Applications	**82**
5. Embedding Social Value in Sustainable Transport and Smart Logistics Systems	**103**
6. Extended Reality in Warehousing: Bridging Industry 4.0 and 5.0 for Smart Logistics	**128**
7. Circular Economy Principles and Sustainable Logistics Practices in Industry 5.0	**149**
8. Reverse Logistics in Industry 5.0: Opportunities and Challenges: A Systematic Literature Review and Research Agenda	**186**
Index	**215**

List of Contributors

Al-Amin Abba Dabo
Department of Business, Systems and Operations, University of Northampton, Northampton, United Kingdom.

Andrea Appolloni
School of Economics, University of Rome, Tor Vergata, 00133 Rome, Italy.

Bright Amajuoyi
Department of Business, Systems and Operations, University of Northampton, Northampton, United Kingdom.

Emmanuel Vanger
Department of Business, Systems and Operations, University of Northampton, Northampton, United Kingdom.

F. Frank Alparslan
Department of Business, Systems and Operations, University of Northampton, Northampton, United Kingdom.

Farideh Moosabeiki
Dehabadii School of Economics, University of Rome, Tor Vergata, 00133 Rome, Italy.

Hao Yu
Department of Industrial Engineering, UiT—The Arctic University of Norway, 8514 Narvik, Norway.

Lin Ma
School of Transportation and Logistics, Southwest Jiaotong University, Chengdu, China.

Mi Gan
School of Transportation and Logistics, Southwest Jiaotong University, Chengdu, China

Mohammadamin Hosseinifard
Department of Civil and Industrial Engineering, University of Pisa, Pisa, Italy.

Qian Zheng
School of Transportation and Logistics, Southwest Jiaotong University, Chengdu, China.

Sina Shahbazi Manshadi
School of Economics, University of Rome, Tor Vergata, 00133 Rome, Italy.

Tao Peng
School of Transportation and Logistics, Southwest Jiaotong University, Chengdu, China.

Wei Deng Solvang
Department of Industrial Engineering, UiT—The Arctic University of Norway, 8514 Narvik, Norway.

Xinchi Dong
School of Transportation and Logistics, Southwest Jiaotong University, Chengdu, China.

Xu Sun
Department of Industrial Engineering, UiT—The Arctic University of Norway, 8514 Narvik, Norway.

Yuxin He
School of Transportation and Logistics, Southwest Jiaotong University, Chengdu, China.

Zihan Zhang
School of Transportation and Logistics, Southwest Jiaotong University, Chengdu, China.

Zijun Guo
School of Transportation and Logistics, Southwest Jiaotong University, Chengdu, China.

1 Distribution and Logistics Network Design in Industry 5.0: Models, Methods, and the Impact of Generative AI and Large Language Models

Hao Yu[1*], Xu Sun[2], Wei Deng Solvang[3] and Mi Gan[4]

1.1 Introduction

Distribution network design is one of the most critical strategic decisions in a logistics system or a supply chain, which seeks to structure the framework by defining the number of echelons and specifying the type, quantity, and location of various facilities (Mangiaracina et al., 2015, Perl and Sirisoponsilp, 1988), and optimizing other mid-term planning decisions such as demand allocation and inventory management (Gupta et al., 2000). The distribution network connects various players and facilities within a logistics system or a supply chain, including production plants, distribution centers (DCs), warehouses, retail stores, customers, as well as other relevant entities (Cordeau et al., 2021). It may also incorporate recycling and disposal facilities when reverse logistics for value recovery at the end-of-life (EOL) stage is considered (Yu and Solvang, 2016). Thus, distribution network design is a complex problem due to the involvement of numerous stakeholders and the need to coordinate various independent decisions at both strategic and tactical/operational levels related to establishing and operating the network, such as the establishment of local production plants, the selection of suppliers and transportation modes, and so forth (Cordeau et al., 2021). As highlighted by Chopra and Meindl (2007), the key factors influencing distribution network design include a company's strategic objectives, technological advancements, political considerations, infrastructure availability, local competition, and overall costs related to facilities, labor, and transportation. In addition, several economic factors, e.g., exchange rates, tariffs, and tax incentives, play a significant role. By effectively managing the interplay among these key influencing factors, distribution network design enables the achievement of a company's or supply chain's strategic objectives, ranging from cost competitiveness to high responsiveness to customer demand (Javid and Azad, 2010) to fostering an environmentally and socially sustainable image (Vafaei et al., 2020).

[1,2,3] Department of Industrial Engineering, UiT—The Arctic University of Norway, 8514 Narvik, Norway.

[4] School of Transportation and Logistics, Southwest Jiaotong University, Chengdu, China.

* Corresponding author: hao.yu@uit.no

DOI: 10.1201/9781003466451-1

Mathematical optimization is the most widely used approach for distribution network design (Klose and Drexl, 2005). Over the past four decades, extensive research has focused on developing and applying mathematical optimization models to tackle diverse distribution network design problems in logistics systems and supply chains (Gendreau, 2023), aiming to simultaneously optimize facility locations, capacity installations, demand allocations, vehicle routing, and inventory levels (Javid and Azad, 2010). Fleischmann et al. (2006) further demonstrated the significant economic benefits that can be achieved through the integrated optimization of decisions across both strategic and tactical/operational levels. Moreover, due to changing market conditions, emerging technologies, increasing global competition, and evolving regulations, distribution and logistics networks must be continuously re-optimized to identify the best ways to adapt to these dynamic circumstances (Cordeau et al., 2021).

The recent Fifth Industrial Revolution, known as Industry 5.0, has drastically transformed the paradigms of many industries (Leng et al., 2022), presenting both opportunities and challenges for distribution and logistics network design. Building on the technology-driven advancements and digitalization in smart manufacturing and logistics spearheaded by its predecessor, Industry 4.0, Industry 5.0 emphasizes human-centricity, sustainability, and resilience in the technology-enabled societal transition (Jafari et al., 2022). On the one hand, distribution and logistics network design must holistically account for not only economic perspectives but also environmental and social objectives to ensure alignment with the triple bottom line of sustainable development. For example, environmental footprints (Dukkanci et al., 2019) and employment generation objectives (Dukkanci et al., 2019) are incorporated alongside economic goals in the recent mathematical optimization models. Furthermore, cutting-edge technologies, e.g., the Internet of Things (IoT), big data analytics, and artificial intelligence (AI), have provided enhanced data capabilities to improve analytical processes and have enabled a high level of methodological integration through digital logistics twins (Ivanov and Dolgui, 2021, Gerlach et al., 2021, Sun et al., 2023). On the other hand, adopting new technologies can also present challenges. For instance, it adds complexity to the initial network design due to a highly dynamic and uncertain environment (Sun et al., 2024). Moreover, issues such as cybersecurity risks, technological maturity, and reliability may act as significant hindrances (Sun et al., 2022).

The purpose of this chapter is to first provide a comprehensive introduction to the general mathematical optimization models and methods for distribution and logistics network design. Subsequently, it discusses the impact of Industry 5.0, with a particular focus on Generative AI (GenAI) and Large Language Models (LLMs). Section 1.2 begins with a continuous network design problem, commonly referred to as the gravity location model or greenfield analysis. Sections 1.3, 1.4, and 1.5 then focus on discrete network design problems, which are more commonly encountered in logistics systems and supply chains. These include capacitated and uncapacitated network design problems, single-source network design problems, and capacitated network design with facility configuration. Numerical examples based on the same case study are provided to demonstrate the application of these mathematical optimization models. Finally, this chapter discusses the impact of GenAI and LLMs in advancing the research and education of distribution and logistics network design. As GenAI and LLMs drive revolutionary changes across numerous fields (Amankwah-Amoah et al., 2024), understanding their advantages and limitations becomes thus paramount.

1.2 Gravity Location Model

The gravity location model, also known as the center-of-gravity model, is an essential analytical tool in distribution and logistics network design to identify optimal facility locations for production plants, warehouses, and DCs. Inspired by Newton's law of universal gravitation (Capoani, 2023), the primary objective is to minimize transportation costs and enhance service efficiency by calculating the geographic "center of gravity" based on the locations of various nodes in the network (e.g., suppliers, customer markets, retail outlets) and the volume of goods to be shipped from/to each (Ballou and Srivastava, 2007). Originating from the location theory from the early 20th century, especially influenced by Alfred Weber's work related to industrial location and transportation cost minimization (Weber, 1922), the gravity location model involves computing the weighted average of the coordinates of all nodes, treating the service area as a continuous plane and employing Euclidean distance matrices to effectively find the "center of gravity" of the distribution network (Revelle and Eiselt, 2005). Therefore, it essentially addresses a continuous facility location problem, serving as an initial screening tool for identifying desirable areas, but it is not intended for final decision-making.

The notations for formulating the gravity location model are first given as follows:

	Sets
N	Set of nodes, indexed by n; These nodes represent various types of entities within a distribution/logistics network that have either input or output flows, such as suppliers, customers, DCs, etc.
	Parameters
Q_n	Quantity of goods shipped from/to node n
C_n	Unit transportation cost from/to node n
x_n, y_n	Location coordinates of node n in the distribution and logistics network
	Decision variables
x_o, y_o	Location coordinates of the optimally selected facility
	Auxiliary variables
d_n	Euclidean distance between the selected facility and node n

The objective function of the gravity location model is given in Eq. (1.1), which minimizes the total transportation cost of the distribution/logistics network.

Eq. (1.2) computes the Euclidean distance between the selected facility and node n. Notably, while d_n is not a decision variable in this model, its value depends on the choice of the facility location, making it an auxiliary variable that facilitates the optimization process. From the operations research perspective, the gravity location model is an unconstrained nonlinear optimization problem.

$$\text{Min transportation cost} = \sum_{n \in N} d_n Q_n C_n \quad \text{...(1.1)}$$

$$d_n = \sqrt{(x_o - x_n)^2 + (y_o - y_n)^2} \quad \text{...(1.2)}$$

Table 1.1 Experiment data

City	*Unit transportation cost (NOK/km/kg)*	*Annual demands (kg)*	*Location*		*Planner coordinates*	
			Latitude	*Longitude*	x_n	y_n
Bergen	1.5	15591	60.39299	5.32415	2639.792	1804.182
Drammen	1.5	6622	59.74389	10.20449	3012.401	1747.927
Frederiksted	1.5	5311	59.2181	10.9298	3063.342	1691.4
Oslo	1.5	42340	59.91273	10.74609	3051.901	1765.108
Stavanger	1.5	8877	58.97005	5.73332	2674.276	1677.401
Trondheim	1.5	10741	63.43049	10.39506	3033.023	2135.609

A numerical example is provided to illustrate the application of the gravity location model in distribution and logistics network design. We considered establishing a distribution network of consumer products in the southern part of Norway, where a DC will be set up to serve six major cities including Oslo, Bergen, Drammen, Frederiksted, Stavanger, and Trondheim. The projected annual demands, unit transportation costs, and the latitude and longitude coordinates for each location are detailed in Table 1.1. To implement the gravity location model, the geographic coordinates (latitude and longitude) of the nodes are converted into planar coordinates using the ETRS89/LAEA Europe projection (EPSG:3035). This projection transforms spherical coordinates into a flat, two-dimensional plane, allowing precise Euclidean distance calculations essential for determining optimal facility locations. Using this data, the optimal location for the DC is determined to minimize the total transportation costs.

The gravity location model is implemented using AMPL (A Mathematical Programming Language), which is a high-level language widely used for describing and solving large-scale optimization and mathematical programming problems. AMPL provides a clear and intuitive syntax that closely mirrors the mathematical notations of the model and supports a vast array of optimization solvers through a unified codebase, making it an excellent choice for solving complex optimization problems. Below is the AMPL code that defines and solves the gravity location model:

```
# Define sets
set node;
# Define parameters
param demand {node};
param transport_cost {node};
param x {node};
param y {node};
# Define decision variables
var optX;
var optY;
# Define the objective function
minimize total_cost:
sum {j in node} transport_cost[j] * demand [j] * (((optX-
x[j])**2 + (optY-y[j])**2)**(1/2));
```

Due to the nonlinear nature of the gravity location problem, the nonlinear solver MINOS is chosen to effectively find the optimal solution. The total transportation cost is calculated to be 21842278 NOK, and the optimal facility coordinates are found to be x_0 =3051.9 and y_0 = 1765.11. These coordinates correspond to the geographic location of Latitude 59.91273° N and Longitude 10.74609° E, which closely aligns with the city of Oslo. This outcome suggests that, given the specific parameters of the model, Oslo is the optimal location for the DC in the distribution network.

While the gravity location model provides a strategic starting point for facility placement by simplifying complex spatial relationships into a manageable calculation, it operates under idealized assumptions (Fotheringham and Wegener, 1999), e.g., flat terrain and Euclidean distances, which hardly capture real-world complexities like geographic obstacles, infrastructure variability, road access, or regulatory constraints. Therefore, it is often used in conjunction with more detailed, discrete location models and practical considerations to inform final site selection in a distribution and logistics network design process.

1.3 Capacitated Network Design Problems

In this section, the capacitated network design problem and its two variations are introduced. Unlike the continuous facility location problem, which aims at pre-screening the entire geographical area, the capacitated network design is a typical discrete facility location/network design problem (Melkote and Daskin, 2001). It focuses on selecting the optimal solutions from a set of predefined discrete locations, enabling more targeted and efficient decision-making in the placement of facilities and the optimization of flows among facilities (Melo et al., 2009).

To model the capacitated network design problem, a simple distribution network design problem based on Chopra and Meindl (2007) is investigated. This network includes two types of nodes: production plants and customers. Each customer's demand must be met by the production plants, respecting their given capacities, and ensuring all customer demands are completely satisfied. In this capacitated setting, demands from one customer can be distributed across multiple production plants, and conversely, a single plant can serve multiple customers. The model's objective is to determine the optimal number and locations of the production plants and the allocation of demand between the plants and customers in a manner that minimizes the overall logistics cost.

The sets, parameters, and decision variables for the capacitated network design problem are first given as follows:

	Sets
I	Set of production plants, indexed by i
J	Set of customers, indexed by j
	Parameters
f_i	Fixed operating cost of production plant i
D_j	Customer demand at location j
c_{ij}	Unit transportation cost between production plant i and customer j
	Decision variables
y_i	Binary decision variable determining if candidate location i is selected to open the production plant
x_{ij}	Demand allocation between production plant i and customer j

The given capacitated network design problem is formulated as follows:

$$\text{Min total cost} = \sum_{i \in I} f_i y_i + \sum_{i \in I} \sum_{j \in J} c_{ij} x_{ij} \qquad \text{...(1.3)}$$

S.t.

$$\sum_{i \in I} x_{ij} = D_j, \forall j \in J \qquad \text{...(1.4)}$$

$$\sum_{j \in J} x_{ij} \leq K_i y_i, \forall j \in I \qquad \text{...(1.5)}$$

$$y_i \in \{0, 1\}, \forall i \in I \qquad \text{...(1.6)}$$

$$x_{ij} \geq 0, \forall i \in I, j \in J \qquad \text{...(1.7)}$$

The optimization model is formulated as a mixed integer linear program, incorporating both integer and continuous variables. It features a typical two-stage decision-making structure: the first stage uses binary integer variables to determine the optimal facility locations, and the second stage employs continuous variables to allocate customer demands to each production plant. From a decision-making perspective, the first-stage facility location decisions are strategic and need to be robust, designed to provide long-term solutions. Conversely, while the model can optimize allocation decisions simultaneously, these second-stage decisions are tactical or operational and can be adjusted later to enhance system performance under various scenarios (Yu and Solvang, 2020). This flexibility allows for responsive adjustments to operational strategies as market conditions evolve.

The objective function, as presented in Eq. (1.3), aims to minimize the total cost of the two-echelon distribution network. The first term of this equation represents the total costs of operating the facilities, while the second term accounts for the transportation costs between production plants and customers. The model is governed by constraints outlined in Eqs. (1.4–1.7). Eq. (1.4) serves as the demand satisfaction constraint, ensuring that the demands at all customer locations are met. Eq. (1.5) represents the capacity constraint, which ensures that the capacities of the selected production plants are not exceeded. Notably, when y_i equals to 0, Eq. (1.5) stipulates that no products will be dispatched from production plant i. Finally, Eqs. (1.6) and (1.7) define the domains of both binary and continuous variables in the model, respectively.

The AMPL code for the capacitated network design problem is given as follows:

```
# Define sets
set Supply; #Set of potential production plants
set Demand; #Set of customers
# Define parameters
param demand {Demand};
param supply {Supply};
param fixed_cost {Supply};
param cost {i in Supply, j in Demand};
# Define decision variables
var quantity {i in Supply, j in Demand} >= 0;
var Y {i in Supply} binary;
# Define objective function
minimize total_cost:
sum{i in Supply} fixed_cost[i]*Y[i] + sum {i in Supply, j
in Demand} cost[i,j]*quantity[i,j];
# Define the supply and demand constraints
subject to supply_constraint {i in Supply}:
sum {j in Demand} quantity[i,j] <= supply[i]*Y[i];
subject to demand_constraint {j in Demand}:
sum {i in Supply} quantity[i,j] = demand[j];
```

A numerical experiment is conducted to demonstrate the application of the capacitated network design model, using the same customer locations as

Table 1.2 Candidate locations of the production plants and distance matrix

Node	*Location*		*Distance matrix*					
	Latitude	*Longitude*	*Oslo*	*Bergen*	*Trondheim*	*Stavanger*	*Drammen*	*Frederiksted*
i1	60.34824	10.722656	85	450	474	477	126	172
i2	60.44088	5.9106445	426	65	623	237	405	518
i3	62.19469	8.1958007	391	486	253	665	432	477
i4	59.84480	9.4042968	117	374	608	391	73	184

Table 1.3 Optimal results of facility selection and demand allocation

Candidate	*Facility selection*	*Demand allocation x_{ij}*					
	y_i	*Oslo*	*Bergen*	*Trondheim*	*Stavanger*	*Drammen*	*Frederiksted*
i1	1	32259		10741			
i2	1		15591		8877		
i3							
i4	1	10081				6622	5311

in the previous case. Following pre-screening by the gravity location model, two candidate locations near Oslo and Drammen are identified, along with two additional candidates considering the customers in other markets. The locations of these four candidates, labeled *i1*, *i2*, *i3*, and *i4*, are detailed in Table 1.2. The distance matrix between the candidate locations and the customers is automatically generated using the Bing Map API, as shown in Table 1.2. The annualized fixed operating costs of the production plants *i1*, *i2*, *i3*, and *i4* are set to 5500000 NOK, 5000000 NOK, 5000000 NOK, and 5500000 NOK, respectively, with each facility having a capacity of 43000 kg/year. The unit transportation cost is set to 1.5 NOK/kg/km, and the customer demands at each market retain the same value as the previous case.

Based on the given data, the optimization problem is solved with the Gurobi solver (version 11.0.2), which is the most powerful commercial solver for mixed integer programs, and the optimal objective value is 36385930 NOK. Table 1.3 presents the decisions of facility selection and demand allocation in the optimal solution. As can be seen, three production plants at locations *i1*, *i2*, and *i4* are selected. Demand from Oslo is distributed between production plants *i1* and *i4*, while demands from other markets are met by individual plants. The capacity utilization at production plants *i1*, *i2*, and *i4* are 94.4%, 56.9%, and 51.2%, respectively.

Two variations of the capacitated network design problem are explored. The first variation is the single-source network design problem, where the demand from each customer must be fulfilled exclusively by one production plant. To formulate this single-source capacitated network design problem, a new decision variable u_{ij} is introduced to replace x_{ij} in the previous model. All other parameters and variables retain their original meanings and functions. This modification ensures that each customer is connected to only one production plant, simplifying the assignment and potentially enhancing the manageability of the distribution and logistics network.

Decision variables	
u_{ij}	Binary decision variable determining if the demand from customer j is assigned to production plant i

The model is then reformulated in Eqs. (1.8–1.11). Notably, the second term in the objective function, Eq. (1.8), has been modified. In this model, u_{ij} represents the selection of demand allocation; therefore, if $u_{ij} = 1$, all the demands from customer j must be satisfied by production plant i. Consequently, D_j (the demand of customer j) needs to be incorporated into the calculation of the transportation cost in order to ensure that the transportation costs are correctly calculated based on the total demand assigned to each production plant. The demand satisfaction constraint is reformulated in Eq. (1.9), which specifies that only one production plant can be chosen to fulfill the demand from one customer. Eqs. (1.10) and (1.11) define the capacity constraint and the domains of the decision variables in the single-source capacitated network design problem, respectively.

$$\text{Min total cost} = \sum_{i \in I} f_i y_i + \sum_{i \in I} \sum_{j \in J} c_{ij} u_{ij} D_j \quad ...(1.8)$$

S.t.

$$\sum_{i \in I} u_{ij} = 1, \forall j \in J \quad ...(1.9)$$

$$\sum_{j \in J} u_{ij} D_j \leq K_i y_i, \forall i \in I \quad ...(1.10)$$

$$f_i, u_{ij} \in \{0, 1\}, \forall i \in I, j \in J \quad ...(1.11)$$

The AMPL code for the single-source network design problem is given as follows:

```
# Define sets
set Supply;
set Demand;
# Define parameters
param demand {Demand};
param supply {Supply};
param fixed_cost {Supply};
param cost {i in Supply, j in Demand};
# Define decision variables
var X {i in Supply, j in Demand} binary;
```

```
var Y {i in Supply} binary;
# Define the objective function
minimize total_cost:
sum {i in Supply} fixed_cost[i]*Y[i] + sum {i in Supply,
j in Demand} cost[i,j]*X[i,j]*demand[j];
# Define supply and demand constraints
subject to supply_constraint {i in Supply}:
sum {j in Demand} X[i,j]*demand[j] <= Y[i]*supply[i];
subject to demand_constraint {j in Demand}:
sum {i in Supply} X[i,j] = 1;
```

Using the same dataset as the previous experiment, the single-source network design problem is now solved. This approach ensures consistency across all experiments in the chapter, facilitating direct comparison of results and evaluating the impacts of the single-source strategy relative to those observed in the earlier model. The optimal objective value for the single-source capacitated network design problem is 37341400 NOK, reflecting a 2.6% increase compared to the standard capacitated network problem. Table 1.4 displays the optimal location-allocation decisions, highlighting that the incorporation of the single-source constraint necessitates the selection of all candidate locations. From an optimization perspective, the single-source constraint narrows down the feasible solution space, typically leading to a less economically efficient solution. This restriction reduces the flexibility in assigning demand across multiple production plants, often leading to increased costs as more locations must be selected to handle dedicated material flows. Furthermore, opening more production plants can lead to underutilization of some facilities. For instance, in this scenario, the newly opened production plant *i3* is operating at only 25% capacity.

Table 1.4 Optimal decisions for the single-source capacitated network design problem

Candidate	*Facility selection* y_i	*Demand allocation*					
		Oslo	*Bergen*	*Trondheim*	*Stavanger*	*Drammen*	*Frederiksted*
i1	1	1					
i2	1		1		1		
i3	1			1			
i4	1					1	1

The second variation is the capacitated network design with facility configurations. In this scenario, multiple capacity alternatives may be installed at the same location, requiring the model to determine not only the facility locations but also the size and capacity installed at each location. This variation adds a layer of complexity as it involves optimizing both the spatial distribution of facilities and their respective capacities to effectively meet demand and operational efficiencies. To model the capacitated network design problem with facility configurations, a new set of capacity alternatives W is defined, tailored to each facility. Utilizing this set, the fixed facility operating costs and the variables for facility opening and demand allocation are redefined as follows:

Sets	
W	Set of the capacity alternatives, indexed by w
Parameters	
f_{iw}	Fixed operating cost of production plant i with capacity w
Decision variables	
y_{iw}	Binary decision variable determining if candidate location i is selected to open the production plant with capacity w
x_{iwj}	Demand allocation between production plant i with capacity w and customer j

The capacitated network design problem with facility configurations is modeled in Eqs. (1.12–1.16). The objective function and constraints are similar to those in the standard capacitated network design outlined in Eqs. (1.3–1.7). The primary distinction lies in the inclusion of set W, which allows the model to also determine the capacity installed at each facility location. Crucially, it is feasible to select more than one configuration at a single facility location if a larger facility configuration can optimize the overall distribution and logistics flows.

$$\text{Min total cost} = \sum_{i \in I} f_{iw} y_{iw} + \sum_{i \in I} \sum_{w \in W} \sum_{j \in J} c_{ij} x_{iwj} \qquad ...(1.12)$$

S.t.

$$\sum_{i \in I} \sum_{w \in W} x_{iwj} = D_j, \forall j \in J \qquad ...(1.13)$$

$$\sum_{j \in J} x_{iwj} = K_{iw}\, y_{iw}, \forall i \in I, w \in W \qquad ...(1.14)$$

$$f_{iw} = \{0, 1\}, \forall i \in I, w \in W \qquad ...(1.15)$$

$$x_{iwj} \geq 0, \forall i \in I, w \in W, j \in J \qquad ...(1.16)$$

The AMPL code for capacitated network design with facility configuration is given as follows:

```
# Define the sets
set Supply; #The set of production plants
set Demand; #The set of customers
set Size; #The set of alternative capacities installed
# Define the parameters
param demand {Demand};
param supply {i in Supply, k in Size;
param fixed_cost {i in Supply, k in Size};
```

```
param cost {i in Supply, j in Demand};
# Define the decision variables
var quantity {i in Supply, k in Size, j in Demand} >= 0;
var Y {i in Supply, k in Size} binary;
# Define the objective function
minimize total_cost:
sum {i in Supply, k in Size} fixed_cost[i,k]*Y[i,k] + sum
{i in Supply, j in Demand} cost[i,j]*(sum {k in Size}
quantity[i,k,j]);
# Define the supply and demand constraints
subject to supply_constraint {i in Supply, k in Size}:
sum {j in Demand} quantity[i,k,j] <= supply[i,k]*Y[i,k];
subject to demand_constraint {j in Demand}:
sum {i in Supply, k in Size} quantity[i,k,j] = demand[j];
```

The numerical experiment is expanded to incorporate two different capacities at each location, namely, high and low. The annualized fixed costs and capacities for the two different capacity levels at each candidate location are detailed in Table 1.5. The optimal total cost of the distribution network, calculated at 31772252 NOK, indicates a substantial cost reduction, achieving a 12.7% decrease compared to the standard capacitated network design problem. Table 1.6 illustrates the optimal decisions related to facility location, capacity installation, and demand allocation. The results reveal that a large production plant at candidate location *i1* has been selected to serve the markets of Oslo, Drammen, and Frederiksted. In addition, two smaller production plants have been placed: one at *i2* to cater to the markets of Bergen and Stavanger, and the other at *i3* to meet the demand from Trondheim. The utilization rates of the production plants *i1*, *i2*, and *i3* are 90.4%, 81.6%, and 35.8%, respectively. These results demonstrate that by employing varying capacity configurations at each location, the overall economic performance of the distribution and logistics network can be enhanced. This improvement is achieved through demand-driven capacity installations at the respective production plants, allowing for more efficient resource use and cost optimization.

Table 1.5 The annualized fixed costs and capacities for both high and low capacity levels at each candidate location

Candidate	*High*		*Low*	
	Fixed cost f_{iw} (NOK)	*Capacity K_{iw} (kg)*	*Fixed cost f_{iw} (NOK)*	*Capacity K_{iw} (kg)*
i1	7000000	60000	4500000	30000
i2	6000000	60000	4000000	30000
i3	6000000	60000	4000000	30000
i4	7000000	60000	4500000	30000

Table 1.6 Optimal results of facility selection, capacity installation, and demand allocation

Candidate	*Facility selection* y_{iw}	*Demand allocation* x_{iwj}					
		Oslo	*Bergen*	*Trondheim*	*Stavanger*	*Drammen*	*Frederiksted*
i1	High	42340				6622	5311
i2	Low		15591		8877		
i3	Low			10741			
i4							

1.4 Uncapacitated Network Design Problems

Although capacitated network design models are widely used in distribution and logistics systems, they are often limited by rigid capacity constraints that can lead to suboptimal recommendations. For example, consider a distribution network design problem where the total customer demand is 1001 units, and the capacities of four potential candidate locations are each 500 units. In this scenario, the capacitated network design model would suggest opening three production plants to ensure all customer demands are met. However, this recommendation is clearly inefficient, as opening an additional production plant to accommodate just one unit of excess demand is disproportionate and economically impractical (Yu and Solvang, 2018).

From a modeling perspective, one approach to address this limitation is to incorporate a penalty cost into the objective function, allowing for minor violations of the demand satisfaction constraint. This adjustment ensures that slight increases in demand do not necessitate opening a new facility; instead, these small excesses are managed through a penalty reflected in the objective function. However, the calibration of the unit penalty cost requires careful consideration. It must be set high enough to deter reliance on the penalty as a cost-saving measure, preventing the model from suggesting a "do-nothing" strategy where all demand is penalized because it is cheaper. On the other hand, if the unit penalty cost is defined as significantly higher, it may become ineffective to limit the opening of another facility (Yu and Solvang, 2018). Incorporating a penalty cost into the model not only helps avoid opening excessive facilities in a distribution and logistics network but also holds practical significance. It recognizes that slight demand increases can be managed through various practices, e.g., overtime work, the employment of part-time or seasonal workers, outsourcing, or even accepting a loss in sales (Yu and Solvang, 2020). These measures typically incur higher costs and are commonly implemented in real-world operations. By incorporating a penalty cost for minor demand overages, the model better aligns with practical business strategies, offering a flexible and economically viable response to small demand fluctuations (Yu and Solvang, 2020).

Next, we consider another solution to this problem, where the capacity of potential production plants is not predetermined, enabling the optimization model to define the size of the plant to be installed. This represents a typical uncapacitated network design problem, where plant capacity is treated as a decision rather than a fixed input (Contreras and Fernández, 2012, Verter, 2011). This adjustment provides enhanced flexibility in the network's response to varying demand levels. The question then arises: How should the uncapacitated network design problem be formulated? Specifically, can we simply eliminate the capacity constraint of Eq. (1.5) from the standard capacitated network design model? While removing this constraint is feasible when $y_i = 1$, complications arise when $y_i = 0$. It's important to note that Eq. (1.5) serves dual functions: not only does it cap facility capacity when $y_i = 1$, but it also prevents dispatch from unselected candidate locations when $y_i = 0$. Simply removing Eq. (1.5) could lead to a situation where no candidate locations are selected, yet products are still dispatched from these unselected locations.

$$\sum_{j \in J} x_{ij} \leq My_i, \forall i \in I \qquad ...(1.17)$$

$$\sum_{j \in J} u_{ij} D_j \leq Mf_i, \forall i \in I \qquad ...(1.18)$$

Thus, to address the issue described earlier, Eqs. (1.17) and (1.18) are introduced to replace Eqs. (1.5) and (1.10) in the standard and single-source capacitated network design problems, respectively. Notably, *M* represents a big number, which is commonly used in optimization models to effectively enforce conditional constraints. The value of *M* requires careful validation. If set too high, *M* can lead to scaling issues within the optimization solver, potentially affecting computational efficiency and accuracy. Conversely, if *M* is too low, it restricts the problem and transforms it into a capacitated network design. Typically, *M* should be slightly larger than the sum of all customer demands. This size ensures it is feasible to open a single large facility capable of fulfilling all customer demands without artificially limiting the solution space.

Table 1.7 Comparison between the capacitated and uncapacitated network design problems under both standard and single source settings

Model	*Capacity*	*Objective value*	*Facility selection*
Standard network design	Capacitated	36385930	*i1, i2, i4*
	Uncapacitated	30832893	*i1, i2*
Single-source network design	Capacitated	37341400	*i1, i2, i3, i4*
	Uncapacitated	30832893	*i1, i2*

Using the same dataset as in the previous section on standard capacitated network design, a numerical experiment is conducted to evaluate the impact of the capacity constraint. In this experiment, the value *M* of is set to 100000, slightly exceeding the sum of the total demands from all customers. Practically, setting *M* at 100000 means there are no restrictions on the capacity of the production plant at any candidate location. Consequently, a single large plant could potentially meet all customer demands if doing so minimizes the overall distribution and logistics costs. Table 1.7 presents the comparison between the capacitated and uncapacitated network design problems under both standard (multi-source) and single-source settings. For the uncapacitated network design problems, both standard and single source settings yield the same results by opening to production plants at candidate locations *i1* and *i2*. Production plant *i1* has a larger annualized capacity at 65014 kg, while production plant *i2* has a smaller annualized capacity at 24468 kg. The demands from Drammen, Frederiksted, Oslo, and Trondheim are allocated to production plant *i1*, while the markets in Bergen and Stavanger are served by production plant *i2*.

As indicated in the table, substantial cost savings are realized in both scenarios. In the standard network design, transitioning to an uncapacitated model results in a 15.3% reduction in costs. Similarly, in the single-source network design, the shift to an uncapacitated framework leads to a 17.4% decrease in total costs. These figures underscore the financial benefits of removing capacity constraints when optimizing distribution and logistics network designs. However, it's important to note that uncapacitated network design may not always be realistic in many real-world situations (Shishebori and Babadi, 2018), where facility size is often constrained by local regulations, physical space limitations, resource availability, or environmental impact considerations.

1.5 The Role of Generative AI and Large Language Models

Today, artificial intelligence (AI) stands out as one of the most promising technologies propelling Industry 5.0, poised to drive disruptive changes across numerous industries and business sectors. Among the latest and most popular AI breakthroughs are Large Language Models (LLMs), which have garnered enormous attention since the release of ChatGPT by the San Francisco–based technology entrepreneur company OpenAI on November 30, 2022 (Dhamani, 2024). Built on advanced natural language processing (NLP) techniques and deep learning (DL) within a generative transformer architecture (Qadir, 2023), ChatGPT, as well as other similar LLMs (e.g., Bard, Gemini, etc.), are trained on vast amounts of data to produce context-relevant, human-like text responses through web-based chatbots and interfaces (Akhavan and Jalali, 2024, Shen et al., 2023). With a focus on text generation, LLMs belong to the broader category of Generative AI (GenAI), which is capable of creating other forms of content including images, music, and videos based on patterns learned from existing data (Akhavan and Jalali, 2024).

GenAI has recently transformed numerous industries and sparked disruptive social changes. In healthcare, for instance, it has been used as an assistant in interactive computer-aided diagnosis, summarizing electronic health records (Shen et al., 2023), and predicting patient hospital stays based on medical records (Yu et al., 2023), among other applications. Furthermore, AI tools have dramatically reshaped the current research landscape by boosting productivity in tasks such as brainstorming, proofreading, coding, and beyond (Stokel-Walker and Van Noorden, 2023). This transformative impact is exemplified by the 2024 Nobel Prize recipients, Geoffrey Hinton and John J. Hopfield, whose pioneering work on enabling machine learning in artificial neural networks (ANNs) laid the foundation for modern AI. Meanwhile, Demis Hassabis and John Jumper from Google DeepMind highlighted the revolutionary potential of AI with AlphaFold, a groundbreaking solution for predicting protein structures that have opened new frontiers in biomedical research. However, the widespread adoption of GenAI tools also raises a range of ethical concerns and challenges to academic integrity. For instance, these models can generate seemingly credible but factually incorrect answers, creating risks around misinformation (Stokel-Walker and Van Noorden, 2023). Moreover, issues such as data protection and unequal access to GenAI across different regions of the world further complicate the responsible use of these technologies (Alasadi and Baiz, 2023).

Education is one of the most critical areas for understanding the impact of GenAI and developing effective response strategies. A recent review indicates that, over the past two years, more than 400 papers have been published on the use of GenAI in education (Yusuf et al., 2024). Of these, over 75% focus on its impact on higher education, and roughly 50% examine its applications, potential, and overall effect (Yusuf et al., 2024). The remaining studies delve into attitudes toward GenAI, ethical guidelines, risks, and both institutional and individual adoption, among other topics (Yusuf et al., 2024). On the one hand, GenAI tools can significantly enhance learning outcomes by effectively supporting both educators and students (Giannakos et al., 2024). For example, educators can use these tools to design classes more efficiently and streamline the preparation of course materials, presentations, codes, and quizzes (Qadir, 2023). Meanwhile, students can benefit from GenAI's capabilities in idea generation, problem-solving, coding, proofreading, and real-time evaluation (Alasadi and Baiz, 2023, Qadir, 2023). One of the most compelling advantages of using GenAI is its ability to support adaptive learning for students, creating a more flexible, distributed, and individualized learning environment without substantially increasing resource requirements (Qadir, 2023, Mittal et al., 2024). However, on the other hand, incorporating GenAI in education also introduces challenges related to data privacy, algorithmic biases, academic integrity, and the potential erosion of human-led critical thinking and creativity (Alasadi and Baiz, 2023). To solve these challenges, Su and Yang (2023) suggested a four-step *IDEE* framework, where the effective integration of GenAI in education needs to consider the following fundamental pillars:

- *I*: The identification of expected learning outcomes
- *D*: The determination of the appropriable level of GenAI involvement
- *E*: The ethical and responsible use of GenAI
- *E*: The evaluation of the effectiveness of GenAI usage

In this section, we demonstrate how GenAI tools can be effectively utilized in both research and education on distribution and logistics network design. Through examples, we examine their advantages as well as potential limitations. To conduct the experiment, ChatGPT o1-mini was used. The subsequent sections examine each of the ten interactions in three parts: **Prompt**, **Response Summary**, and **Comments**, with ChatGPT's responses displayed in italics.

Prompt 1

I am a manager of a logistics system and I want to choose the right number and locations for my production plants. The locations of the four candidates are given and the markets are located in Oslo, Bergen, Trondheim, Stavanger, Drammen, and Frederiksted. How can I determine the number and best locations of the plants? Reply to it in short.

Explain the steps without using mathematical formulas.

Response Summary

Choosing the best among four plant sites is a classic logistics problem, usually tackled with location-allocation models. Here's the short approach:

1. *Collect Your Information:*
   - *You have four places where you might build plants.*
   - *You have six markets (Oslo, Bergen, Trondheim, Stavanger, Drammen, Frederiksted).*
   - *You know roughly how much it costs to open each plant, how much it costs to ship goods from each plant to each market, and how much each market needs.*
2. *Decide on the Variables:*
   - *For each candidate site, ask: "Do I open a plant here, yes or no?"*
   - *For each market, figure out "How much supply does this place get from each open plant?"*
3. *Make a Plan to Minimize Total Cost:*
   - *Total cost = The cost to open plants + The shipping cost to get products to each market.*
   - *You want to choose which plants to open and where each market gets supplied from so that your total cost is as low as possible.*
4. *Follow the Rules (Constraints):*
   - *Each market must receive all of its demand.*
   - *You can't ship goods from a place if you didn't open a plant there.*
   - *If plants have capacity limits, don't exceed those when assigning demands.*
5. *Use a Solver:*
   - *Input all of the above into a solver (specialized software or even a spreadsheet tool).*
   - *The solver will figure out the best solution—telling you which plants to open and how to supply each market.*

6. *Read the Results:*
   - *You'll see which candidate sites should be opened (minimizing cost).*
   - *You'll see how much each open plant ships to each market.*

Comments

The responses from ChatGPT provide an introduction for individuals without a relevant background. While the initial explanations included mathematical formulas, requesting a more straightforward, formula-free explanation resulted in more accessible and easy-to-understand responses. The procedures are correct and clearly articulated, making them ideal for students to begin their learning journey.

Prompt 2

Can you help define the mathematical models to solve this problem?

Response Summary

Two alternative mathematical models are recommended and the respective formulations are provided accordingly and explained:

- *Uncapacitated facility location model: the model decides which plants to open and how much each open plant will supply to each market to minimize total costs.*
- *P-median model: You fix the number of plants to open, and the model chooses which sites to open and which market is served by which open site, minimizing total cost/distance.*

Comments

ChatGPT provided two alternative models to solve the problem. Although the uncapacitated facility location problem may be applicable to the given scenario, the *p*-median model is generally employed for service facility location problems that prioritize accessibility (Albareda-Sambola et al., 2015). Furthermore, other relevant alternatives, such as the capacitated facility location problem and the gravity model, were not suggested. Therefore, in terms of model selection, ChatGPT may not offer solutions tailored to different scenarios and might propose inappropriate models.

In terms of model formulation, ChatGPT provided accurate mathematical models for both problems and explained their key components, enabling readers to thoroughly understand the foundational structures and assumptions inherent in each model.

Prompt 3

Should the *p*-median model be more relevant for a service facility location problem than this distribution network design?

Response Summary

The p-median model is more common in service facility problems (e.g., deciding where to place fire stations or hospitals). There, you typically want to minimize travel distance/time to each customer or population area, and you often have a fixed number of facilities you can open.

In distribution network design, you often care about both fixed costs (opening plants) and transportation costs (shipping to markets). The Uncapacitated Facility Location model captures these costs directly and lets you decide how many facilities to open based on cost trade-offs.

So, if you only know you want to open a fixed number of facilities and you measure distance rather than cost, p-median is indeed more aligned with service facility location. But if your problem focuses on cost and doesn't strictly fix the number of sites, the facility location model is more typical.

Comments

To explore this issue, we asked ChatGPT why the *p*-median model was proposed for solving a distribution network design problem, considering it is generally used for service facility locations. ChatGPT clarified the distinctions between the two facility location models and outlined their appropriate application scenarios. It concluded that selecting the most suitable model requires careful consideration of the specific circumstances within the decision-making process.

Prompt 4

What if the capacity or the number of facilities to be opened has already been decided?

Response Summary

If you already know the number of plants to open and you have capacity limitations, you can combine the ideas from both the p-median and facility location models:

- *Fixed Number of Plants by adding the constraint* $\sum_{j \in J} y_j = p$*, which enforces exactly p open plants.*
- *Capacity Constraints (if needed) by adding* $\sum_{i \in I} x_{ij} \leq Cap_j \cdot y_j, \forall j \in J$*. This says plant j can supply the demand of multiple markets, but no more than its capacity.*

Comments

In this scenario, we intentionally posed a question to ChatGPT with the aim of directing it to generate a capacitated network design problem. ChatGPT responded by offering two alternative solutions, both of which effectively address and accommodate the specific practical situations presented. Consequently, when consulting ChatGPT for model selection, outlining the conditions in detail is necessary to receive a proper answer. Notably, when addressing our questions, ChatGPT employed varying notations, designating I as the set of customers and J as the set of candidate locations for plants.

Prompt 5

What if a capacitated facility location problem should be modeled?

Response Summary

Below is a Capacitated Facility Location Problem (CFLP) formulation in concise terms:

...

$$x_{ij} < d_i \cdot y_i, \forall i \in I$$

Shipping only from open facilities: if a facility is closed ($y_i = 0$), it cannot ship goods. (This is a typical "big-M" constraint, ensuring no shipments occur from a facility that isn't open)

Comments

We proceeded by asking ChatGPT to formulate a capacitated facility location problem, resulting in a well-designed model, as detailed in Section 1.3. Notably, ChatGPT added an extra constraint to ensure that shipping is conducted only from open facilities. Nevertheless, this constraint is redundant, as the capacity constraint already ensures this condition when the binary decision variable is set to 0, as discussed in Section 1.3.

Prompt 6

Is the constraint of shipping only from open facilities a redundant one? Its function can be achieved with the capacity constraint in the capacitated facility location problem. Why do you give a redundant constraint in this model?

Response Summary

In many Capacitated Facility Location formulations, the "no shipping from closed facilities" constraint can become redundant if you already have a tight capacity constraint of the form. In uncapacitated versions (or if capacities are extremely large and don't effectively limit shipments), you often do need a direct "no shipping if closed" constraint to ensure no flow from a closed facility. In practice, many standard textbooks and formulations include that "no shipping from closed facilities" constraint for clarity, so, it's often kept in the model for completeness, clarity, or to handle large-capacity (or uncapacitated) cases.

Comments

We continued to question ChatGPT about the inclusion of a redundant constraint in the model. ChatGPT conceded that the constraint was indeed redundant and offered reasons for its presence. However, the provided justification was not entirely persuasive. For example, although the constraint enforcing shipping exclusively from open facilities is prevalent in many textbooks, it is usually paired with a simplified capacity constraint, defined as $\sum_{j \in J} u_{ij} D_j \leq K_i, \forall i \in I$. In such cases, the capacity constraint is particularly relevant when $y_i = 1$ (indicating an open facility), while the shipping constraint specifically handles situations where $y_i = 0$ (indicating a closed facility). Furthermore, the concept of "extremely large capacity" does not apply when $y_i = 0$, as no shipments should originate from a closed facility. Consequently, ChatGPT's explanation may be misleading, as it fails to adequately address the distinct roles of these constraints and the binary variable in the model.

Prompt 7

Can you generate the AMPL code for the capacitated facility location model? Can you generate the respective data file? Can you convert the model with single source constraints and to uncapacitated facility location problems?

Response Summary

The model file is as follows:

```
set I; # Set of markets
set J; # Set of candidate facility locations
param d{i in I}; # Demand of market i
param c{i in I, j in J}; # Unit shipping cost from facility
j to market i
param f{j in J}; # Fixed cost to open facility j
param Cap{j in J}; # Capacity of facility j (finite)
var y{j in J} binary; # 1 if facility j is opened, 0
otherwise
var x{i in I, j in J} >= 0; # Quantity shipped from
facility j to market i
# Objective: Minimize total fixed + shipping cost
minimize Total_Cost:
 sum {j in J} (f[j] * y[j])
 + sum {i in I, j in J} (c[i,j] * x[i,j]);
# Constraint: Demand satisfaction
subject to Demand_Constraint {i in I}:
 sum {j in J} x[i,j] = d[i];
# Constraint: Capacity limitation
subject to Capacity_Constraint {j in J}:
```

```
sum {i in I} x[i,j] <= Cap[j] * y[j];
```

The data file is as follows:

```
#-----------------------------------------------------
# Annual demand (kg) for each market
#-----------------------------------------------------
param d :=
 Oslo           42340
 Bergen         15591
 Trondheim      10741
 Stavanger      8877
 Drammen        6622
 Frederiksted   5311;
#---------------------------------------------------
# Fixed annual operating costs (NOK)
#-----------------------------------------------------
param f :=
 i1  5500000
 i2  5000000
 i3  5000000
 i4  5500000;
#-----------------------------------------------------
# Capacity (kg/year) for each candidate location
#-----------------------------------------------------
param Cap :=
 i1  43000
 i2  43000
 i3  43000
 i4  43000;
#-----------------------------------------------------
# Transportation cost matrix: c[i, j] in NOK/kg
# (Distance * 1.5 NOK/kg/km)
#-----------------------------------------------------
param c :
              i1      i2      i3      i4 :=
Oslo          127.5   639.0   586.5   175.5
Bergen        675.0   97.5    729.0   561.0
Trondheim     711.0   934.5   379.5   912.0
Stavanger     715.5   355.5   997.5   586.5
Drammen       189.0   607.5   648.0   109.5
Frederiksted  258.0   777.0   715.5   276.0;
#-----------------------------------------------------
```

The codes were then properly converted to respective variations

Comments

We asked ChatGPT to code the mathematical model for our study. Initially, ChatGPT generated the model file, after which we provided the relevant data outlined in Section 3. First, we supplied the location information and requested the creation of a distance matrix. ChatGPT responded by calculating the Euclidean distances based on the provided coordinates. Subsequently, we asked it to generate a distance matrix using real-world distances, to which ChatGPT recommended employing a professional GIS system for accurate measurements.

To test its data handling capabilities, we deliberately removed one dataset and asked ChatGPT to identify the missing data. It successfully pinpointed the missing dataset required for model implementation. Ultimately, both the model file and the data file were generated. Upon validation, we confirmed that ChatGPT produced the correct code for the given problem and data.

Furthermore, we instructed ChatGPT to expand the model by incorporating a single-source requirement and relaxing the capacity constraints. ChatGPT adeptly adjusted both the model and data files accordingly. The only issue observed was the occasional redundancy of the additional constraint requiring shipping exclusively from open facilities.

Prompt 8

If we consider CFLP as a benchmark and compare the two variations including the UFLP and single-source CFLP, which variation can yield a better solution? And why?

Response Summary

The Uncapacitated Facility Location Problem (UFLP) generally yields better, more cost-effective solutions compared to the Single-Source Capacitated Facility Location Problem (Single-Source CFLP) when using CFLP as a benchmark.

- *UFLP allows each market to receive products from multiple facilities without capacity limits, providing greater flexibility and enabling more efficient, lower-cost allocations.*
- *Single-Source CFLP restricts each market to be served by only one facility and imposes capacity constraints, which can lead to higher costs and less optimal resource utilization.*

In summary, UFLP's flexibility in assignments and absence of capacity constraints typically result in more optimal and economical solutions than Single-Source CFLP.

Comments

Next, we tasked ChatGPT with comparing the three various models to ascertain which would provide the best results. ChatGPT delivered well-supported answers and clear rationales for its assessments.

Prompt 9

What are the limitations of a capacitated facility location problem?

Response Summary

Some common limitations of the Capacitated Facility Location Problem (CFLP) include:

- *Complexity and Scalability: CFLP is NP-hard, so it becomes computationally expensive for large-scale problems.*
- *Static Assumptions: It generally treats the problem as single-period and ignores changing conditions over time.*
- *Accuracy of Data: The model's effectiveness depends on precise and reliable input data.*
- *Simplification of Real Conditions: It overlooks complexities like multiple products and service-level requirements.*
- *Limited Focus on Sustainability: Environmental or social impacts are typically not accounted for.*
- *Dependency on Predefined Parameters: Results rely heavily on estimates for costs, capacities, and demands.*
- *Interdependencies and non-linearities: Real-world logistics can include non-linear interactions not captured by a linear model.*

Comments

Addressing this question is particularly difficult because the flexibility issue inherent in the capacitated facility location problem is often overlooked in standard textbooks. ChatGPT's responses covered areas such as model complexity, underlying assumptions, data uncertainty, and the application of linear models to real-world scenarios. On one hand, these aspects are pertinent to the capacitated facility location problem, providing valuable insights into its formulation and application. On the other hand, the discussions were quite general and could be applicable to a wide range of similar optimization problems. Crucially, ChatGPT did not identify or discuss the flexibility issue, which is the most significant limitation of the capacitated facility location model. This omission highlights a gap in the GenAI's analysis, as the lack of flexibility can severely impact decision-making in dynamic environments.

Prompt 10

We are now considering the adoption of electric trucks for transportation within the logistics network. Can the CFLP be further developed to simultaneously determine the facility location of plants and the location of charging stations? The electric trucks can be charged at markets and plants, and the newly installed charging stations only need to ensure the en-route charging between different nodes in the network.

Response Summary

Below is the mathematical formulation of the enhanced Capacitated Facility Location Problem (CFLP) that simultaneously determines the locations of production plants and charging stations for electric trucks within the logistics network. In this model, electric trucks can be charged at both plants and markets and charging stations are also installed en-route to ensure that trucks can recharge between different nodes within their range (R). Herein, a new set of K is defined for the candidate locations of charging stations, and d_{ik} and d_{jk} are the distance between the charging stations and market i and plant j, respectively.

The following constraints are then defined:

- *Plant Charging Coverage: Ensure that each opened plant has at least one charging station within the truck's range R*

$$\sum_{k \in K: d_{jk} \leq R} z_k \geq y_j, \forall j \in J$$

- *Market Charging Coverage: Ensure that each market has at least one charging station within the truck's range R*

$$\sum_{k \in K: d_{ik} \leq R} z_k \geq 1, \forall j \in J$$

- *En-route Charging Coverage: To ensure en-route charging, impose that any shipment route from a plant to a market can be serviced by existing charging stations. A simplified way is to ensure that for each shipment from Plant j to Market i, there exists at least one charging station k such that:*

$$d_{jk} + d_{ik} \leq 2R$$

This implies that the trip from Plant j to Charging Station k and then to Market i does not exceed twice the truck's range, allowing for a recharge at k. To integrate this into the model, introduce the following constraint:

$$\sum_{k \in K: d_{ik} + d_{jk} \leq 2R} z_k \geq 1, \forall i \in I, j \in J,$$

In addition, the constraints for CFLP are incorporated in the model.

Comments

Finally, we challenged ChatGPT with a research question aimed at establishing a joint logistics and charging infrastructure network, similar to Yu and Sun (2024). In response, ChatGPT proposed a solution that considered both node coverage and en-route charging. Specifically, it correctly formulated the plant charging coverage and market charging coverage. To address the en-route charging requirement, ChatGPT introduced a constraint ensuring that at least one charging station is opened between each established pair of plant and market nodes. However, this constraint is fundamentally flawed.

Firstly, in scenarios where the distance between two nodes significantly exceeds twice the charging radius (i.e., much longer than $2R$), the constraint becomes infeasible, necessitating the opening of multiple charging stations to maintain connectivity. Conversely, when two nodes are in close proximity (i.e., much shorter than R), the requirement to open a charging station between them is redundant and unnecessary. Consequently, the constraint proposed by ChatGPT does not adequately account for varying distances between nodes, rendering the solution incorrect. Nevertheless, the proposed solution by ChatGPT can serve as a valuable starting point for further discussion and problem-solving by highlighting key considerations in infrastructure network design.

Based on our experiments, we find that GenAI is an effective tool for enhancing research and teaching in distribution and logistics network design. Specifically, it assists and significantly enhances the following aspects:

- *Well-organized introduction to existing problems*: GenAI excels in synthesizing and presenting comprehensive introductions to current challenges in distribution and logistics network design, providing a clear and structured overview that facilitates quick understanding.
- *Model selection particularly with detailed information*: When provided with detailed information, GenAI aids in selecting the most appropriate models by analyzing the specific requirements and nuances of the problem, ensuring that the chosen models align well with the given data and objectives.
- *Model formulation and explanation to enhance learning effectiveness*: GenAI effectively formulates complex models and offers clear and detailed explanations of their components and underlying principles, thereby improving the learning effectiveness of students.
- *Significantly reduced coding time and improved code explanations*: By generating accurate and efficient code for various models, GenAI drastically reduces the time required for coding tasks. Besides, it provides clear explanations of the code, enhancing users' understanding of the logic and structure behind the implementations.
- *Model comparisons with different features*: GenAI facilitates the comparison of different models by highlighting their various features, strengths, and limitations. This comparative analysis is crucial for selecting the most suitable model for specific scenarios.
- *Idea generation for solving new and emerging research challenges*: GenAI can help generate initial research ideas to solve emerging challenges in distribution and logistics network design.

Despite its numerous advantages, our experiments also revealed several limitations of GenAI in the research and education for distribution and logistics network design:

- *Reliability issues and accuracy*: Our experiments demonstrate that, in numerous instances, ChatGPT provides unreliable or incorrect answers to posed questions. When users do not possess a clear understanding of the underlying problems, these inaccuracies can be highly misleading, potentially hindering the learning process.

- *Lack of critical thinking:* While ChatGPT can generate satisfactory responses to existing problems and effectively utilize textbook knowledge, it fundamentally lacks the capacity for critical thinking that extends beyond established information. The AI tends to regurgitate known solutions with the limited ability to analyze, evaluate, or innovate beyond the provided data.
- *Inability to address emerging challenges:* Although ChatGPT excels in formulating, coding, and solving established models, our tests indicate that it struggles with new and emerging challenges that require the integration of specific conditions and constraints. When faced with novel problems that demand highly innovative and customized solutions, the GenAI often falls short, unable to adapt its responses to accommodate unique or complex requirements.

These limitations may hinder the development of deep analytical and problem-solving skills and diminish effectiveness in flexible, advanced research settings. A recent study has also argued that GenAI and LLMs are incapable of genuine logical reasoning, and they only *replicate reasoning steps from their training data* (Mirzadeh et al., 2024). Furthermore, other general concerns, such as data privacy and inequality (Alasadi and Baiz, 2023), must be considered when implementing GenAI in the research and education of distribution and logistics network design.

1.6 Conclusions

The distribution network determines the long-term performance of a logistics system and is considered one of the most important strategic decisions. Based on the domain of feasible solutions, optimization models can be categorized into continuous facility location models and discrete network design models. First, the gravity location model, a continuous facility location model that explores the entire solution space, is introduced. However, due to its lack of feasibility considerations for the selected locations, the gravity location model can only serve as an initial screening tool rather than for final decision-making. Next, the capacitated network design problem, a discrete network design problem with a solution space consisting of a set of predefined candidates, is presented. This model is further explored through two variations: one with single-source requirements and the other with different facility configurations. Subsequently, the uncapacitated network design problem is introduced. Through numerical experiments based on the same case study, these models are thoroughly compared to illustrate their differences and application scenarios.

Finally, we discuss the impact of GenAI and LLMs on the research and education of distribution and logistics network design. In our experiments, we designed and tested a set of questions ranging from introductory to highly research-focused levels using ChatGPT o1-mini. Our results reveal that GenAI tools can serve as effective assistants in solving distribution and logistics network design problems, thereby drastically improving research productivity and learning outcomes. Specifically, GenAI excels in providing an introductory framework of existing knowledge, explaining and coding models for existing problems, and facilitating model selection and comparison.

However, researchers, educators, and learners must also clearly understand the limitations of current GenAI tools. For example, when asking a GenAI tool to select a model, detailed contextual information must be provided to avoid inappropriate recommendations. Furthermore, while GenAI tools are valuable for initial idea generation and brainstorming, they are less effective for solving cutting-edge research challenges due to their lack of genuine critical thinking and inability to address emerging problems that extend beyond the training data, as demonstrated in our experiments. Therefore, when adopting GenAI in research and education, its role must be clearly defined, and its effectiveness should be evaluated using the *IDEE* framework.

Nevertheless, one limitation of our experiments is that the conclusions are drawn within the context of distribution and logistics network design, and the applicability of GenAI may vary across different domains.

Declaration of Generative AI and AI-assisted Technology in the Writing Process

During the preparation of the manuscript, ChatGPT o1-mini was used for the experiments in Section 5 to evaluate the GenAI tools in distribution and logistics network design. Besides, the authors used ChatUiT (a tailored version of ChatGPT) to polish the scientific writing. After using these tools, the authors reviewed and edited the content as needed and take full responsibility for the content of the paper.

Acknowledgment

The research is supported by the Industry 5.0 enabled Smart Logistics: A Global Perspective project funded by the Norwegian Directorate for Higher Education and Skills (HK-dir) under the Utforsk Programme (UTF-2021/10166), the UiT Aurora project MASCOT, and the International Cooperation Plan under the Department of Science and Technology in Sichuan Province (No.2024YFHZ0315).

References

Akhavan, A. and Jalali, M.S. 2024. Generative AI and simulation modeling: how should you (not) use large language models like ChatGPT. *System Dynamics Review*, 40, e1773.

Alasadi, E. A. and Baiz, C.R. 2023. Generative AI in education and research: Opportunities, concerns, and solutions. *Journal of Chemical Education,* 100, 2965–2971.

Albareda-Sambola, M., Hinojosa, Y. and Puerto, J. 2015. The reliable p-median problem with at-facility service. *European Journal of Operational Research,* 245, 656–666.

Amankwah-Amoah, J., Abdalla, S., Mogaji, E., Elbanna, A. and Dwivedi, Y.K. 2024. The impending disruption of creative industries by generative AI: Opportunities, challenges, and research agenda. *International Journal of Information Management*, 79, 102759.

Ballou, R.H. and Srivastava, S.K. 2007. *Business logistics/supply chain management: planning, organizing, and controlling the supply chain*, Pearson Education India.

Capoani, L. 2023. Review of the gravity model: Origins and critical analysis of its theoretical development. *SN Business & Economics,* 3, 95.

Chopra, S. and Meindl, P. 2007. *Supply chain management. Strategy, planning & operation*, Springer.

Contreras, I. and Fernández, E. 2012. General network design: A unified view of combined location and network design problems. *European Journal of Operational Research,* 219, 680–697.

Cordeau, J.-F., Klibi, W. and Nickel, S. 2021. Logistics network design. *Network design with applications to transportation and logistics*, 599–625.

Dhamani, N. 2024. *Introduction to Generative AI*, Simon and Schuster.

Dukkanci, O., Bektaş, T. and Kara, B.Y. 2019. Green network design problems. *In Sustainable transportation and smart logistics* (pp. 169–206). Elsevier.

Fleischmann, B., Ferber, S. and Henrich, P. 2006. Strategic planning of BMW's global production network. *Interfaces,* 36, 194–208.

Fotheringham, S. and Wegener, M. 1999. *Spatial models and GIS: New and potential models*, CRC press.

Gendreau, M. 2023. Network Design with Applications to Transportation and Logistics.

Gerlach, B., Zarnitz, S., Nitsche, B. and Straube, F. 2021. Digital supply chain Twins—Conceptual clarification, use cases and benefits. *Logistics*, 5, 86.

Giannakos, M., Azevedo, R., Brusilovsky, P., Cukurova, M., Dimitriadis, Y., Hernandez-Leo, D., Järvelä, S., Mavrikis, M. and Rienties, B. 2024. The promise and challenges of generative AI in education. *Behaviour & Information Technology*, 1–27.

Gupta, A., Maranas, C.D. and Mcdonald, C. M. 2000. Mid-term supply chain planning under demand uncertainty: customer demand satisfaction and inventory management. *Computers & Chemical Engineering*, 24, 2613–2621.

Ivanov, D. and Dolgui, A. 2021. A digital supply chain twin for managing the disruption risks and resilience in the era of Industry 4.0. *Production Planning & Control*, 32, 775–788.

Jafari, N., Azarian, M. and Yu, H. 2022. Moving from Industry 4.0 to Industry 5.0: what are the implications for smart logistics? *Logistics*, 6, 26.

Javid, A. A. and Azad, N. 2010. Incorporating location, routing and inventory decisions in supply chain network design. *Transportation Research Part E: Logistics and Transportation Review*, 46, 582–597.

Klose, A. and Drexl, A. 2005. Facility location models for distribution system design. *European journal of operational research*, 162, 4–29.

Leng, J., Sha, W., Wang, B., Zheng, P., Zhuang, C., Liu, Q., Wuest, T., Mourtzis, D. and Wang, L. 2022. Industry 5.0: Prospect and retrospect. *Journal of Manufacturing Systems*, 65, 279–295.

Mangiaracina, R., Song, G. and Perego, A. 2015. Distribution network design: a literature review and a research agenda. *International Journal of Physical Distribution & Logistics Management,* 45, 506–531.

Melkote, S. and Daskin, M.S. 2001. Capacitated facility location/network design problems. *European journal of operational research,* 129, 481–495.

Melo, M. T., Nickel, S. and Saldanha-Da-Gama, F. 2009. Facility location and supply chain management–A review. *European journal of operational research,* 196, 401–412.

Mirzadeh, I., Alizadeh, K., Shahrokhi, H., Tuzel, O., Bengio, S. and Farajtabar, M. 2024. Gsm-symbolic: Understanding the limitations of mathematical reasoning in large language models. *arXiv preprint arXiv:2410.05229.*

Mittal, U., Sai, S. and Chamola, V. 2024. A comprehensive review on generative ai for education. *IEEE Access.*

Perl, J. and Sirisoponsilp, S. 1988. Distribution networks: facility location, transportation and inventory. *International Journal of Physical Distribution & Materials Management*, 18, 18–26.

Qadir, J. Engineering education in the era of ChatGPT: Promises and pitfalls of generative AI for education. 2023 IEEE Global Engineering Education Conference (EDUCON), 2023. IEEE, 1–9.

Revelle, C.S. and Eiselt, H.A. 2005. Location analysis: A synthesis and survey. *European journal of operational research*, 165, 1–19.

Shen, Y., Heacock, L., Elias, J., Hentel, K.D., Reig, B., Shih, G. and Moy, L. 2023. ChatGPT and other large language models are double-edged swords. Radiological Society of North America.

Shishebori, D. and Babadi, A.Y. 2018. Designing a capacitated multi-configuration logistics network under disturbances and parameter uncertainty: a real-world case of a drug supply chain. *Journal of Industrial Engineering International*, 14, 65–85.

Stokel-Walker, C. and Van Noorden, R. 2023. What ChatGPT and generative AI mean for science. *Nature*, 614, 214–216.

Su, J. and Yang, W. 2023. Unlocking the power of ChatGPT: A framework for applying generative AI in education. *ECNU Review of Education*, 6, 355–366.

Sun, X., Yu, H. and Solvang, W.D. A Digital Reverse Logistics Twin for Improving Sustainability in Industry 5.0. IFIP International Conference on Advances in Production Management Systems, 2023. Springer, 273–286.

Sun, X., Yu, H., Solvang, W.D. and Govindan, K. 2024. A two-level decision-support framework for reverse logistics network design considering technology transformation in Industry 4.0: a case study in Norway. *The International Journal of Advanced Manufacturing Technology*, 134(1), 389–413.

Sun, X., Yu, H., Solvang, W.D., Wang, Y. and Wang, K. 2022. The application of Industry 4.0 technologies in sustainable logistics: a systematic literature review (2012–2020) to explore future research opportunities. *Environmental Science and Pollution Research*, 29, 9560–9591.

Vafaei, A., Yaghoubi, S., Tajik, J. and Barzinpour, F. 2020. Designing a sustainable multi-channel supply chain distribution network: A case study. *Journal of Cleaner Production*, 251, 119628.

Verter, V. 2011. Uncapacitated and capacitated facility location problems. *Foundations of location analysis*, 25–37.

Weber, A. 1922. Über den *Standort der Industrien,* JCB Mohr.

Yu, H. and Solvang, W.D. 2016. A general reverse logistics network design model for product reuse and recycling with environmental considerations. *The International Journal of Advanced Manufacturing Technology,* 87, 2693–2711.

Yu, H. and Solvang, W.D. 2018. Incorporating flexible capacity in the planning of a multi-product multi-echelon sustainable reverse logistics network under uncertainty. *Journal of cleaner production*, 198, 285–303.

Yu, H. and Solvang, W.D. 2020. A fuzzy-stochastic multi-objective model for sustainable planning of a closed-loop supply chain considering mixed uncertainty and network flexibility. *Journal of Cleaner Production*, 266, 121702.

Yu, H. and Sun, X. 2024. Can an Industry-Led infrastructure development strategy facilitate electric truck Adoption? *Transportation Research Part D: Transport and Environment*, 134, 104356.

Yu, P., Xu, H., Hu, X. and Deng, C., 2023. Leveraging generative AI and large language models: a comprehensive roadmap for healthcare integration. *Healthcare*, 11(20), 2776.

Yusuf, A., Pervin, N., Román-González, M. and Noor, N.M. 2024. Generative AI in education and research: A systematic mapping review. *Review of Education*, 12, e3489.

2 Distribution and Logistics Network Design Using Digital Twin: Concepts, Methods, and Applications

Xu Sun[1*], Hao Yu[2], Wei Deng Solvang[3] and Mi Gan[4]

2.1 Introduction

Today, the intensity of market competition has escalated considerably, fueled by diverse consumer preferences, stringent environmental regulations, and global competitors. An essential factor for organizations to not only survive but also thrive in today's competitive environment is their supply chain and logistics performance (Farahani et al., 2014). The distribution network design in logistics systems is a combination of facility- and network-related decisions, involving the determination of the locations of facilities required to meet a set of customer demands, as well as determining the best path links that connect different players in the process (Ghaderi and Jabalameli, 2013).

Initially, the notion of "logistics" was intricately associated with military operations (Serrano et al., 2023). In this framework, logistics involves the strategic planning, implementation, and management of the efficient and effective movement and storage of personnel, equipment, and supplies from their origin to their designated points of use (Ailawadi and Singh, 2011). The primary aim is to ensure that military units receive the necessary support, resources, and services at the exact time and location required to successfully carry out their missions (Kane, 2012). As time has advanced, the core principles of military logistics have been adapted and incorporated into the fields of business and manufacturing, leading to the modern understanding of logistics within supply chain management. Logistics involves the integration of material management and distribution (Rushton et al., 2022). Logistics, in a general business context, refers to the management of the movement of goods, services, and associated information from the point of origin to the point of consumption to fulfill the needs of customers or corporations (Lummus et al., 2001). It encompasses the planning, execution, and control of both forward and reverse flows, as well as the storage of these resources, ensuring efficiency and effectiveness in meeting customer requirements. More specifically, logistics stands as one of the most essential functions in a supply chain, adeptly

[1,2,3] Department of Industrial Engineering, UiT—The Arctic University of Norway, 8514 Narvik, Norway.
[4] School of Transportation and Logistics, Southwest Jiaotong University, Chengdu, China.
* Corresponding author: xu.sun@uit.no

DOI: 10.1201/9781003466451-2

connecting a network of organizations, individuals, activities, information, and resources to ensure the efficient and effective movement of products or services from suppliers to end customers (Van Der Valk et al., 2022).

In recent years, supply chain and logistics management has undergone a paradigm shift, with its focus evolving from traditional operational efficiency to enhanced effectiveness, sustainability, responsiveness, and digitization. This transformation reflects a growing awareness of the impact of logistics activities on the utilization of advanced technologies and environmental awareness, as well as the necessity for governments and enterprises to adopt more convenient and sustainable practices to maintain competitiveness and meet regulatory requirements (Seuring and Gold, 2013). The concept of Industry 4.0 aims to optimize and automate production and logistics through the intelligent interconnection of machines and processes, thereby simultaneously enhancing responsiveness, cost-efficiency, and productivity of the manufacturing industry as well as many other business sectors (Butt, 2020). This concept highlights the integration of advanced technologies, i.e., the Internet of Things (IoT), big data analytics, artificial intelligence (AI), machine learning (ML), smart robots, and virtual technologies to optimize processes, products, and services. The three main characteristics of Industry 4.0 are connectivity, information transparency, and decentralized decision-making (Jacoby and Usländer, 2020). The emerging concept of Industry 5.0 extends Industry 4.0 by emphasizing a human-centric transformation blueprint, focusing primarily on sustainable, social, and environmental dimensions. Within the context of Industry 5.0, the application of cutting-edge core technologies is further advanced, prompting enterprises to transform their business strategies to achieve sustainability (Bhandal et al., 2022, Sun et al., 2023a, Jafari et al., 2022).

Digital twin is one of the most emerging technological enablers and is rapidly gaining traction across many different industries as a quintessential solution to help companies achieve the benchmarks set by Industry 5.0 (Barata and Kayser, 2024). Digital twin is not merely a simulation model (Van Der Valk et al., 2022); it is an integrated system that connects various analytical models, data, tools, and other relevant elements to support real-time data exchange, informed decision-support, and effective system control. A digital twin can be defined as a comprehensive, multifunctional platform that mirrors a system by integrating various physical and network components. These components are linked and coordinated to operate collaboratively, functioning together as a cohesive entity (Sun et al., 2023a, Sun et al., 2022). Therefore, integrating a fully functional digital twin is crucial for enhancing the efficiency, effectiveness, and visibility of a company's logistics systems, thereby enabling better decision support. Moreover, recent progress in AI, big data analytics, and advanced data processing and management tools has markedly enhanced the feasibility of digital twins and their impact on decision support systems. Emerging as a significant and innovative trend, digital twins are utilized across numerous applications. A digital twin is a virtual representation of a real-life entity, empowered by continuous learning and updates from multiple data streams, which supports real-time control and improved decision-making processes (Sun et al., 2023a).

However, selecting the right analytical tools and appropriate data to address specific distribution network and logistics management issues is of utmost importance and requires domain expertise. As shown in Figure 2.1, the integrated system of an intelligent supply chain and logistics management framework combines various AI algorithms, optimization models, and simulations. These elements support corresponding strategic, tactical, and operational decisions based on different problems and are organized into three sub-systems: data collection, data processing and analysis, and decision support. Despite the significant advantages of predictive, prescriptive, and descriptive analytics, the integration of these methods has not been fully developed and faces numerous challenges in system integration. These challenges stem from disparate data sources, incompatible software packages, varying coding languages, different data structures, and complexities associated with data aggregation (Sun et al., 2022). An in-depth exploration of potential applications and key modeling technologies in Industry 5.0 reveals that equipment serves as the foundational infrastructure for industrial scenarios. The embedded intelligent enhancement of equipment is a crucial prerequisite for the development of digital twins (Lv, 2023).

The chapter is organized as follows. Section 2.2 provides a comprehensive background on the conceptual and methodological development of digital twins. Section 2.3 discusses the methods and applications of digital twins in logistics and supply chain management, establishing the foundation for a digital logistics and supply chain twin. Section 2.4 presents numerical experiments on the utilization of a digital logistics twin for distribution network design, comparing these experiments with those in Chapter 1. Finally, Section 2.5 provides the concluding remarks.

2.2 Conceptual and Methodological Development of Digital Twin

Digital twins are among the most promising technologies in Industry 5.0, yet their role as primary technological enablers remains largely underappreciated and underutilized (Lv, 2023, Barata and Kayser, 2024). The concept was initially introduced in the aerospace and aviation sectors during the 1970s when NASA launched Apollo 13 (Li et al., 2021). NASA's initial objective was to devise a method for monitoring spacecraft operations, managing risks, and responding to emergencies, ultimately saving astronauts' lives (Li et al., 2021). Therefore, NASA built a "twin" model to simulate and assess the conditions aboard Apollo 13, enabling both the astronauts and the control center to remotely monitor the spacecraft's status and make critical decisions during emergencies (Bhandal et al., 2022). In 2003, Michael Grieves from the University of Michigan characterized the digital twin as the "digital counterpart to a physical product", and the origins of the digital twin concept are based on the model named "Conceptual Ideal for PLM" (Product Lifecycle Management) (Grieves and Vickers, 2017). This model

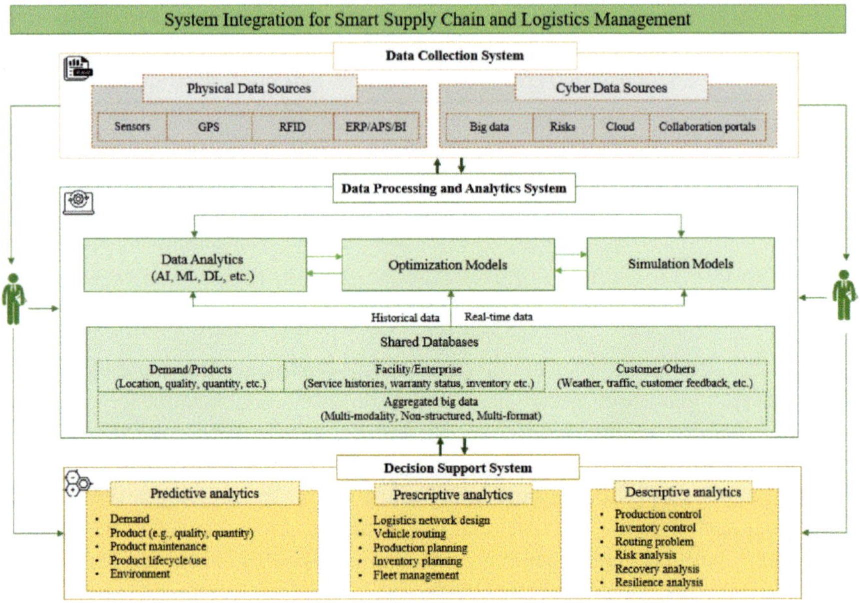

Fig. 2.1 System integration for smart supply chain and logistics management.

contains all the elements of the digital twin: real space, virtual space, and the link for data flow within the physical and digital systems. During that period, another similar concept, 'Digital Shadow,' was introduced. Due to the limitation of communication technologies, the digital shadow results in a unidirectional data flow of the digital shadow, only from the physical entity to the virtual model, which cannot establish a closed data loop (Fuller et al., 2020). Driven by the exponential growth and rapid advancements in technologies like machine learning, wireless communication, and cloud computing, the concept of digital twins is continually evolving, leading to a swift increase in related research activities (Moshood et al., 2021). Digital twins are initially recognized as sophisticated information systems. The driving force behind their development is to create comprehensive asset information visualization strategies specifically designed for asset-heavy industries (Li et al., 2021). Across various academic sources, the digital twin is extensively described as a virtual model of a physical asset. For example, Professor Grieves (Grieves and Vickers, 2017), who first introduced the concept of the digital twin as:

- "*The digital twin is a set of virtual information constructs that fully describes a potential or actual physical manufactured product from the micro atomic level to the macro geometrical level. At its optimum, any information that could be obtained from inspecting a physical manufactured product can be obtained from its Digital Twin*" (Grieves and Vickers, 2017).

Digital twins have been conceptualized from multiple perspectives and within various disciplines, resulting in a wide range of definitions. The scope of digital twins differs considerably due to their diverse applications. Predominantly, most definitions emphasize a specific object, including parts, machines, processes, humans, or physical manufacturing systems. Below are two illustrative examples:

- "*A digital twin is a dynamic and self-evolving digital/virtual model or simulation of a real-life subject or object (part, machine, process, human, etc.) representing the exact state of its physical twin at any given point of time via exchanging the real-time data as well as keeping the historical data*" (Singh et al., 2021).
- "*The digital twin model is an exact and real-time cyber copy of a physical manufacturing system that truly represents all of its functionalities*" (Leng et al., 2019).

As attention to digital twins continues to grow in both academia and industry, their potential to provide processed and valuable information and data is increasingly recognized, thereby enhancing decision-making support. Furthermore, the critical role of different types of analytical models in information processing has been predominantly focused on. Numerous research articles highlight the data analysis capabilities of the digital twin, for instance, other definitions of the digital supply chain twin and digital reverse logistics twin are given by Van Der Valk et al. (2022) and Sun et al. (2023a), respectively:

- "*A digital twin is a virtual representation of a physical artifact (a Supply Chain) that contains a bi-directional data flow between the virtual and digital representation. It provides solutions for further data handling, i.e., simulation, optimization, or monitoring*" (Van Der Valk et al., 2022).
- "*Digital reverse logistics twin is a data-based digital representation of a real-world reverse logistics system, which forms a multi-architecture and high-level integrated information platform by integrating different stakeholders, data, and analytical tools to support various proactive and/or reactive decisions*" (Sun et al., 2023a).

Digital twins hold significant application value in Industry 5.0, making them a rapidly emerging topic. The literature on digital twins is expanding, examining their applications across diverse fields. Digital twins were initially developed and used in the manufacturing industry (Mandolla et al., 2019). The implementation of digital twins is particularly significant in smart manufacturing. For instance, digital twins empower companies to develop virtual manufacturing processes, enabling them to design, test, and identify defects before the actual production process commences. This proactive approach not only enhances product quality but also reduces costs and accelerates time-to-market by minimizing the need for physical prototypes and extensive trial-and-error testing (Schleich et al., 2017). Moreover, when integrated with service-oriented systems, e.g., enterprise resource planning (ERP) and supply chain management (SCM) platforms, digital twins can enhance manufacturing planning and enable precise production control (Qi et al., 2018, Kamble et al., 2022). Stark et al. (2019) investigated the methodological, technological, operational, and business aspects of developing and operating digital twins by using a smart factory cell as a test environment to demonstrate and evaluate the application of smart manufacturing. The DT 8-dimension model offers a framework encompassing the key dimensions of a digital twin solution. It is designed to establish a functional understanding, logical structure, and intended behavioral interactions among physical system elements, software-controlled systems, and data-driven analytic design components. A substantial number of articles explore the applications of digital twin technology across various fields from a broader perspective, most published research tends to concentrate on specific aspects of digital twins, e.g., service (Chen et al., 2024, Rabah et al., 2022, Kobayashi et al., 2023), visibility (Moshood et al., 2021), human-machine interfaces (Mo et al., 2023, Wang et al., 2023), automation (Wilhelm et al., 2020), traceability (Liu et al., 2021), etc. Meanwhile, several studies discuss and investigate the integration of digital twins with specific technologies, e.g., the IoT, blockchain, big data, and additive manufacturing (Tao et al., 2018, Jesus et al., 2024, Liu et al., 2023, Shen and Li, 2024, Qi and Tao, 2018, Figueiras et al., 2021). These discussions focus on the opportunities and challenges of the application and integration of the disruptive technologies involved, thereby providing references and theoretical support for the further development of digital twins. Yet, there is still a gap in methodologies concerning the development of system-oriented digital twins (Fuller et al., 2020), most studies often aimed at developing independent customized applications and isolated operational processes. Furthermore, a comprehensive understanding of digital twins as a core technological element within the entire system, particularly from a supply chain and logistics management perspective, remains scarce (Ivanov, 2024, Park et al., 2020).

2.3 Digital Twin for Logistics and Supply Chain Management

In recent years, logistics and supply chain management have undergone a paradigm shift, evolving from a focus on traditional operational efficiency to priorities of enhanced effectiveness, sustainability, responsiveness, and, most importantly, digitization. This transformation highlights an increasing recognition of how logistics activities influence the integration of state-of-the-art technologies and environmental sustainability. In addition, it emphasizes the need for companies to adopt sustainable practices, fostering environmental consciousness and social responsibility in their supply chain and logistics planning and operations to remain competitive and meet regulatory standards (Seuring and Gold, 2013, Sun et al., 2021). Meanwhile, the swift progress of information and communication technology (ICT) and digitalization in Industry 5.0 has unlocked new opportunities for improving distribution and logistics management. This progress is realized through the adoption of transformative technologies, i.e., IoT, AI, big data analytics, simulation, blockchain, and advanced data processing and management tools, thereby rendering processes more digital, intelligent, sustainable, and efficient (Sun et al., 2023a, Kamble et al., 2020, Bhandal et al., 2022).

Digital twins are among the most prominent technologies in Industry 5.0, exhibiting a wide range of applications within logistics and supply chain management across various phases of the product life cycle, including product design, manufacturing and remanufacturing processes, service operations, and end-of-life (EOL) management (Tao et al., 2018, Preut et al., 2021). For example, Ivanov (2023) introduced a comprehensive decision-making framework that employed digital twins for stress-testing and resilience analysis in supply chains, emphasizing digital twins can contribute to the advancement of supply chain resilience and viability theory. The proposed intelligent digital twin (iDT) system strengthens supply chain resilience by digitally visualizing physical supply chains, processing data, emulating human decision-making, and promoting human-AI collaboration to facilitate monitoring, disruption prediction, and proactive responses. A blockchain-based digital twin solution to re-engineer logistics management systems was proposed by Gai et al. (2022), which enhanced the digitization and intelligence of logistics management to accommodate large service volumes within complex and interconnected industrial systems. This study develops a novel algorithm that significantly outperforms three baseline algorithms, namely, effective iterative greedy (EIG), two dimensional genetic (TDG), and high-level task scheduling dynamic programming (HTSDP), by reducing consensus time and energy consumption by up to 50.05% and 48.46%, respectively, and decreasing the time for generating a consensus plan by a factor of 9.88. Another paper by Maheshwari et al. (2023) shows that digital twins significantly boost food supply chain productivity by enhancing makespan time, reducing data redundancy, optimizing scheduling plans, and improving overall operations and equipment effectiveness, as well as capacity utilization, ultimately achieving a high service level of 94% through seamless integration of production process data. It is noted that there is a wealth of high-quality literature on digital twins, but it is inconsistently scattered and primarily remains at a basic conceptual level. To further demonstrate the significance of digital twins in logistics and supply chain management, more real-world case studies are necessary. For instance, Jesus et al. (2024) illustrated the application of a digital logistics twin in the context of battery recycling to highlight its potential usefulness and practical aspects.

The primary functions of digital twins in logistics and supply chain management include enhancing collaboration and communication, driving value creation, enabling timely decision-making, and promoting sustainability (Bhandal et al., 2022, Ivanov et al., 2019, Moshood et al., 2021, Kamble et al., 2022).

- *Enhanced collaboration and communication*: Digital twins provide a shared and accurate view of logistics systems, enhancing communication and collaboration among stakeholders. This alignment fosters more coordinated and efficient logistics operations.
- *Value creation:* Four types of value including economic, environmental, information, and customer are identified based on a comprehensive review by Schenkel et al. (2015). By simulating and analyzing supply chain processes, digital twins identify opportunities for innovation and efficiency improvements,

leading to cost savings, enhanced service levels, and the creation of new business models that add value for organizations and their customers.

- *Timely and robust decision-making:* Digital twins enable organizations to make informed and robust decisions swiftly by providing historical and real-time data, predictive insights, and simulations, thereby, enhancing resilience and operations, facilitating proactive disruptions management, and responding to demand fluctuations (Ivanov and Dolgui, 2021).
- *Sustainability:* Digital twins support sustainability initiatives by enabling organizations to model, test, and assess the environmental impacts of operational policies in transportation and logistics. This facilitates the identification of strategies to reduce carbon emissions, optimize resources, and ensure compliance with environmental regulations (Rigó et al., 2024).

According to Ivanov et al. (2019), digital twins in logistics and supply chain management enhance decision-making by replicating physical logistics systems or supply chains using real-time data on transportation, inventory, demand, and capacity. This replication enables effective planning and control decisions. By integrating simulation, optimization, and data analytics, a comprehensive technology stack is established, resulting in a digital twin that continuously mirrors the network's real-time state. Furthermore, a digital twin for logistics and supply chain management leverages digital technologies to monitor and influence a system in real-time, while also supporting risk management efforts.

Figure 2.2 illustrates the conceptual framework of the digital twin for logistics and supply chain management. As shown, the main elements include:

- *Physical layer:* Represents the actual entities in a logistics system, e.g., manufacturing plants, warehouses, DCs, customers, trucks, etc.
- *Cyber-physical layer:* Connects entities between the physical and cyber layers through IoT and cyber-physical systems (CPS) interfaces, enabling real-time data collection and synchronization.
- *Cyber layer:* Comprises data analytics, optimization, and simulation tools to process, predict, analyze, and optimize decision-making and operations in logistics systems using historical and real-time data.
- *User interface:* Provides a platform for user-system interaction, offering input windows and output windows for system visualization and performance evaluation.

From a system-oriented perspective, the digital twin for logistics and supply chain management is a virtual system that reflects the current state of the physical system (Ivanov, 2024). Utilizing smart cyber-physical system (CPS) interfaces, digital twins consolidate pertinent data including actual demand, transportation, inventory, and capacity to construct predictive analytics, optimization, and simulation models. This consolidation allows for automated, real-time process analysis across interconnected machines, software, and data sources. Furthermore, by evaluating different scenarios, digital twins expedite decision analyses, improve error detection and correction, and support both proactive and reactive planning (Lv, 2023).

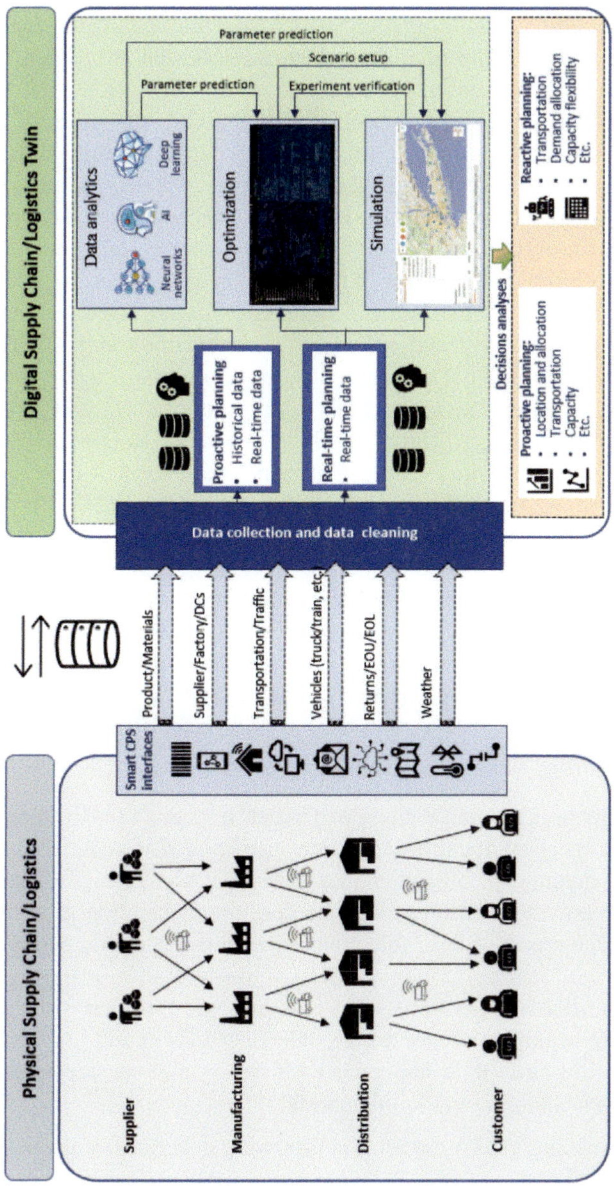

Fig. 2.2 Digital logistics and supply chain twin.

2.4 Digital Logistics Twin for Distribution Network Design

This section presents numerical experiments illustrating the application of digital logistics twins in solving distribution network design problems. Currently, there are several solutions for building highly integrated and intelligent digital logistics twins. Based on the scope of distribution network design, which emphasizes the integration of optimization models and system simulations, the anyLogistix package is utilized in the experiments (Ivanov, 2024, Yu and Sun, 2024). It is important to note that the current solution does not represent a fully functional digital logistics twin, as it excludes the integration of a real physical system and the cyber-physical interconnection. However, it serves as a partial reflection, effectively demonstrating the key conceptual architecture for system integration, as well as the primary components of the cyber layer and user interface.

The anyLogistix system is a professional software package renowned for its versatility in designing and enhancing distribution and logistics networks through various experimental scenarios that deliver optimal, customized solutions. anyLogistix stands out due to its high system integration, merging analytical optimization methods (traditional logistics network design techniques) with dynamic simulation within a single platform, positioning it as a sophisticated tool for logistics and supply chain management (Alx, 2024). Furthermore, it offers users an intuitive interface to interact with numerous pre-built elements of logistics and supply chain systems, including models, outputs, business graphics, and GIS-based maps. By harnessing anyLogistix's powerful integration of optimization and simulation, comprehensive distribution and logistics networks can be efficiently constructed, tested, and evaluated across diverse real-world scenarios and operational conditions.

2.4.1 Greenfield Analysis (GFA) Experiment

Greenfield analysis is an analytical type of experiment that uses the gravity location model to find the best locations for new facilities that will satisfy the given transportation demand of all nodes in a network (Alx, 2024). As introduced in Chapter 1, the gravity location model is a continuous location problem designed for the initial screening of desirable areas. Due to the dynamic nature of demand and supply, as well as the significant impact of transportation costs, location issues are always prioritized. Therefore, GFA is a strategic tool that provides valuable insights to help organizations optimize their distribution and logistics networks. This optimization aims to enhance cost efficiency and service effectiveness in response to ever-changing market demands.

To start building a GFA model, the following data needs considered:

- Products and unit
- Customer demand
- Supply sources
- Customer locations

- Service distance constraints (optional)
- Number of potential facilities (optional)

Note: Optional data is only required when determining the optimal number of facilities within a specified service distance.

Distribution network design implies finding the best locations for new facilities (warehouses, manufacturing plants, etc.), defining the flows between them, and balancing associated costs. In essence, its primary objective is to minimize transportation costs, calculated as the product of distance, unit transportation cost, and the number of products transported.

Following this, a numerical example is presented to showcase the development of a basic GFA model using anyLogistix for a distribution and logistics network design problem. In this chapter, we revisit the case study presented in Chapter 1 to maintain consistency throughout the book. The scenario involves establishing a distribution network for consumer products in the southern part of Norway, where a distribution center (DC) will be set up to serve six major cities: Oslo, Bergen, Drammen, Frederiksted, Stavanger, and Trondheim. The projected annual demands, unit transportation costs, and the latitude and longitude coordinates for each location are detailed in Table 2.1, utilizing the same data as shown in Table 1.1 in Chapter 1. Using this data, the optimal location for the DC is determined to minimize total transportation costs.

Table 2.1 Data for the GFA experiment

City	*Unit transportation cost (NOK/km/kg)*	*Annual demands (kg)*	*Location*	
			Latitude	*Longitude*
Bergen	1.5	15591	60.39299	5.32415
Drammen	1.5	6622	59.74389	10.20449
Frederiksted	1.5	5311	59.2181	10.9298
Oslo	1.5	42340	59.91273	10.74609
Stavanger	1.5	8877	58.97005	5.73332
Trondheim	1.5	10741	63.43049	10.39506

Font description in modeling:

In the modeling process, **Bold** indicates the name of the list of actions for the selected modeling element in the table editor or experiment under selected scenarios. *Italic* is used for the input values.

Modelling steps:

a. Create a GFA scenario.

First, click the **Create Scenario** button on the welcome screen. In the dialog box that appears, name the scenario *GFA*; the scenario type is already set to GFA by default. Set both the **Start date** and **End date** to *1/1/25* and *12/31/25* (default setting: one year), which will define the experiment's running time when the GFA model is running. For **Description,** adding some descriptions or leaving it blank is optional.

Next, enabling the **Advanced settings** button, a larger scenario dialog box will open, which is a data generation box. Within this dialog box, the proper data generation type needs to be chosen by select country/countries, demand generation type, number of locations, number of products, etc. Based on the experimental data provided earlier, as shown in Figure 2.3, **Select country/countries**: *Norway;* the data generation type is chosen: *Demand proportional to population density* which means one customer per location with demand proportional to population density; **Number of locations**: *6;* **Demand per citizen per period**: *0.001,* **Order placement period (days)**: *5*; and also by enabling the **Create country groups for customers** button, ALX can create different customer group by country which provides easier and quicker way for the following modeling steps; and then click **Create**.

In addition, the six customers can be manually added via the GIS map. Begin by selecting the customer icon (blue) and placing six random customers by double-clicking on the map. Next, navigate to **Locations**, paste the **Latitude** and **Longitude** values from Table 2.1 into the editing table, and rename the corresponding locations.

b. Adding parameters to the GFA model.

Once the new GFA scenario has been created, the experimental data needs to be filled in. Because the data generation box is used in advanced settings, the demand will be automatically filled in the table. However, if the customer locations are added manually, the demands also need to be added manually.

Manually add customer demand, first, locate and click on **Demand** in the lower-left corner, then select *Historical Demand* from the **Demand Type** dropdown list in the lower-right editing interface, the last step is filling in the total demands for all the customers under the **Parameters.** Manually entering data is not recommended, especially when creating larger models, as it can be very inefficient and prone to errors. Another way is to use the Generate button in **Demand**. Once clicking the **Generate** button (at the bottom), a little box will open, change the **Demand Type** to *Historic Demand*, set **Order interval, days** to *5*, and **Quantity** is set to *0.001 per citizen*, then the demand will be automatically calculated and added to the table as well.

Other basic data need to be set as well. Select **Products,** the **Unit** is set to *kg*.

c. Run the GFA experiment.

There are two primary types of GFA experiments. The first type assumes that all routes are based on Euclidean distance, disregarding roads, geographical features, and other real-world constraints. While this approach simplifies the model, it may yield infeasible solutions, as discussed in Chapter 1. The second type, known as **GFA with roads experiment**, incorporates actual road networks and geographical considerations, thereby generating more accurate and feasible results.

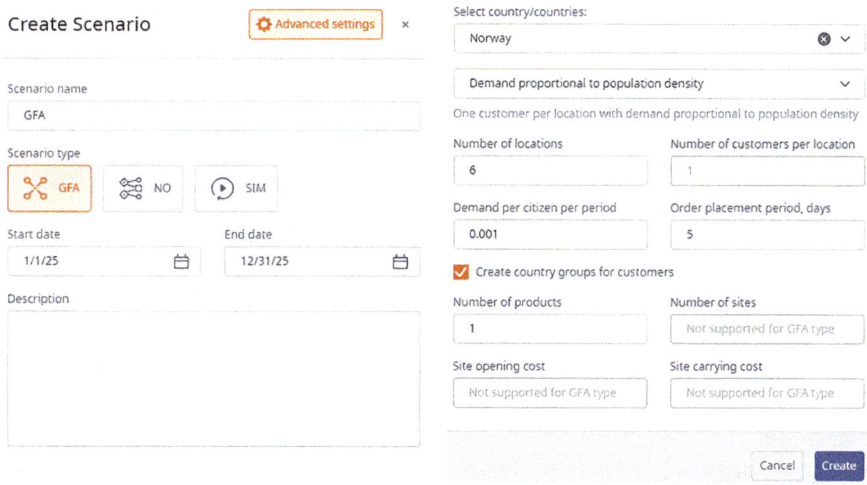

Fig. 2.3 Advanced settings for data generation in anyLogistix.

Running the experiment is also very simple, just by setting some parameters, for instance, in this numerical experiment, the **Number of sites** is set to *1*; the **Product stats unit** is set to *kg;* the **Distance stats unit** is set to *km;* **Destinations** represent the groups of objects which need to be selected correctly, in this simple GFA model, only customers need to be considered, so enabling *(All customers)* button and the others can be set as default or change based on needs.

The last step is to run the model by clicking the **Run** button. When the experiment is finished, a new result item is created in the GFA experiment branch. By clicking the **Result**, several tables are shown in different ways, and the results we obtained can be further used and analyzed in distribution and logistics network design. The analysis determined the optimal location for the *new GFA DC* to be at a latitude of *59.913* and a longitude of *10.746,* which is the same result obtained in Chapter 1. As illustrated in the map view, the new DC is situated near Oslo.

Furthermore, as can be seen, another option in the GFA experiment settings is called **Service Distance**. When service distance constraints are applied and the experiment is executed, the number of potential facilities within the specified service distance that can meet customer demand in a timely manner and improve service level performance is easily determined.

2.4.2 Network Optimization (NO)

Distribution and logistics network optimization focuses on identifying the most effective configuration of different stakeholders within the system. The primary objective is to align supply with demand while minimizing overall costs. A network optimization model describes the distribution network as a set of all possible flows and facilities. Conducting a network optimization experiment in anyLogistix aims to determine the best network design configuration (Alx, 2024). This process seeks to find the most efficient way to transport products from supply nodes to receiving nodes, taking into account various logistics constraints. The experiment will yield several optimal or near-optimal network configuration options, including the one with the lowest cost. By leveraging the results of the network optimization, a logistics manager can evaluate different distribution network designs and determine the potential economic performance of each configuration in order to select the most suitable option.

To build anyLogistix network optimization model, the following information needs to be considered:

- Demand
- Locations of suppliers, DCs, customers, existing and other potential facilities
- Product flows
- Product storages
- Products and unit
- Path (Transportation cost)
- Facility expenses
- Vehicle types
- Assets constraints
- Periods

These data are essential for constructing the NO model in anyLogistix. The software incorporates the IBM ILOG CPLEX® solver (Alx, 2024). Furthermore, constraints play a crucial role in anyLogistix by assisting in the determination of feasible solutions. When setting parameters for network optimization experiments, various constraints can be employed, including demand constraints, product flow constraints, inventory capacity constraints, and production constraints (Alx, 2024). However, achieving optimal results that satisfy all constraints is not always feasible, and infeasibility or errors may arise due to rigid constraints. In such cases, converting some hard constraints into soft constraints can be considered, as discussed in Chapter 1. In anyLogistix, corresponding penalties can be set, incurring penalties for processes that violate constraints. This approach facilitates the generation of more feasible network optimization solutions.

2.4.3 Capacitated Network Design Problems in AnyLogistix

Next, the anyLogistix Network Optimization (NO) model is utilized to address capacitated network design problems. Employing the same numerical study presented in Chapter 1 and based on the data in Table 1.2, the NO model can be constructed through the following steps.

Modelling steps:

a. Create a GFA scenario.

To build a NO experiment, the locations of customers and factories, as well as customer demand, must be set. The previous steps outlined in the GFA can be followed, utilizing the same method to generate the input demand data for NO.

First, click the **Create Scenario** under Network Optimization, enable **Advanced setting** button, and select the data generation type: *Demand proportional to population density (one customer per location with demand proportional to population density)*; **Country**: *Norway;* enabling the **Create country groups for customers** button; **Number of locations**: *6;* **Demand per citizen per period**: *0.001,* **Order placement period(days)**: *5;* **Number of products**: *1*, the others are set to *0*, and click **Create**. Then, the demands for 6 customers were generated automatically in this NO model.

b. Adding and editing parameters to the NO model

b.1 Edit the demand, change **Periodic demand** to **Historic demand**. Click the **Demand,** go to the **Demand Type**, select *Historical demand* from the drop-down list, and fill in the corresponding value of different customer demands one by one manually. The second way is using the **Generate** button, changing the demand type to *Historic demand*, setting **Order interval, days** to *5*, and **Quantity** to *0.001* per citizen, then the demand will be calculated and filled in the table automatically, then delete the previous demands (because it will generate new demands below). Both ways can get the same setting.

b.2 Currency is set to *NOK*. Ensure that all currency units in this NO model are changed to *NOK*. The currency unit needs to be kept consistent, otherwise an error will be reported when the experiment is run. (Note: Press the Shift button and click the **Currency** column, next click the Space button, then enable edit the whole selected cells at one time). If there is not a *NOK* option, one new unit needs to be added in **Manage units** by clicking the icon in the upper right corner.

b.3 Add 4 factories. Select create factory icon (orange icon) and add 4 random factories by double-clicking on the map, then go to **Locations,** paste the **Latitude** and **Longitude** values from Table 1.2 into the editing table, and rename the corresponding locations as *Factory Oslo, Factory Bergen, Factory near Trondheim,* and *Factory Drammen,* respectively. Then, the same map view including both customer and factory locations will be obtained, as shown in Figure 2.4.

b.4 Create groups. The group function is very useful for the following steps, which can help you set up the models in a quick way. Find the **Groups** option in the editing menu bar in the lower left corner, and click **Add** button at the bottom. Add one customer group which includes all 6 customers and add another factory group which includes 4 factories.

b.5 Add facility expenses. Go to **Facility Expenses**, click **Add**, and choose the group of factories *[Factories]* from the drop-down list under the **Facility**. It is evident that each production plant incurs a fixed cost when the facility is operational. In anyLogistix, the initial cost is utilized to represent this fixed cost. Consequently, the **Initial Cost** under **Expense Type** is selected and set to *5500000 NOK* for this group of factories. After all the settings have been done, next, click the factories group *[Factories]*, and click on **Split Rows** button at the bottom,

Column(s) to split within the selected row(s): *Facility*, by clicking **Split Rows,** so, it can break it down into four factories respectively. Last, change the fixed cost (**Initial Cost)** of Bergen and Near Trondheim to *5000000 NOK*. This is the simple and quicker way to set the parameters considering if you have 100 facilities in your network design, otherwise, the data needs to be added and edited one by one.

b.6 Add one supplier. Select the **Create Supplier icon** (green icon), then double-click on the map to add a supplier.

b.7 Add two paths

(1) Path 1: *Supplier location → [Factories]*, choose *Fixed delivery* of **Cost Calculation**, set *0* to **Cost unit** in the **Cost Calculation Parameters,** this is because the transportation cost from the supplier to the factories is not considered. Due to the limitation of the software, the supplier must be included to make this model run and be able to model the capacity constraints in a proper way. In addition, the *LTL (Less truck load)* needs to be chosen as a **Transportation Policy**.

(2) Path 2: *[Factories] → [Customers]*, choose *Product & Distance-based* of **Cost Calculation**, set *kg* to **Amounts unit** and *1.5* to **Cost per unit** in the **Cost Calculation Parameters**, and choose *LTL* as **Transportation Policy**, see Figure 2.5 below.

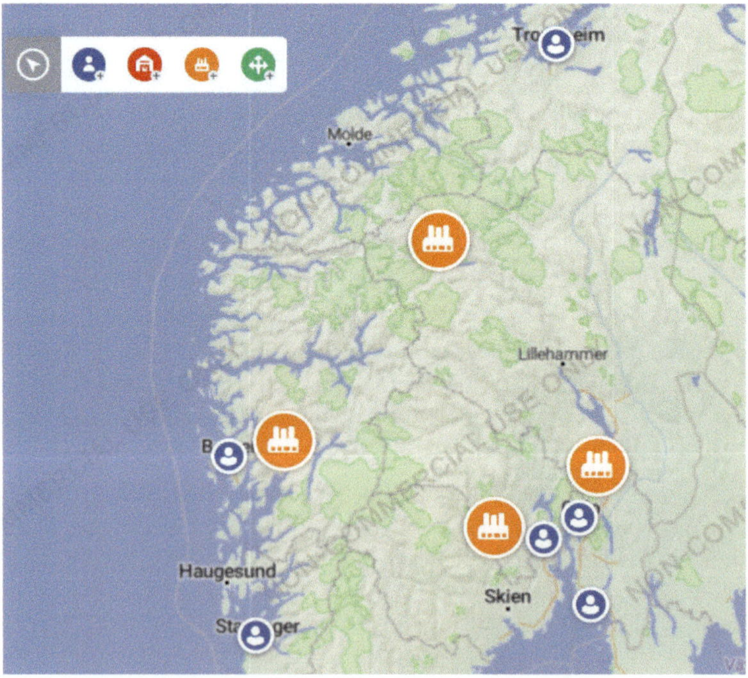

Fig. 2.4 The locations for both candidate facilities and customers.

b.8 Add product flows, and product flows enable the configuration of sourcing parameters, specifying the movement of products from designated sources to their respective destinations.

(1) Flow 1: *Supplier → [Factories]*, Flow 2: *[Factories] → [Customers].* Considering the capacity constraints of each factory, the **Max Throughput** is set to *43000* for both product flows. The default settings enable the **Expand Sources/Destinations** option, which ensures that the capacity of each facility is 43000 for each product.
(2) **Currency** sets to *NOK.*

b.9 Edit **Product Storages**. Considering the capacity constraints of each factory, the **Max Stock** is set to *43000* for all factories. By default settings, 0 in stock setting which means the stock capacity is unlimited. Because the penalties are set to 0 by default, it means that this is a hard constraint and cannot be violated. It is worth noting that under hard conditions, this may lead to infeasible solutions. On the contrary, if the penalties are greater than 0, it is a soft constraint. The solution is obtained by paying the corresponding penalties. Nevertheless, the value of the penalty cost needs to be thoroughly evaluated, as discussed in Chapter 1.

b.10 Edit **Product**, change the **Unit** to *kg;* **Currency** sets to *NOK.* The product cost is not taken into account, so both **Selling price** and **Cost** are set to *0.*

b.11 **Vehicle types** can be renamed or left by default. Set the **Capacity unit** to *kg* and **Capacity** of the vehicle to *0,* which means it is unlimited capacity. In this model, the vehicle capacity is not considered.

b.12 DCs and Factories, we need to *disable* the **Initially Open** toggle button and select *Consider* as the **Inclusion type**. By configuring the model in this manner, it autonomously determines which facilities should remain operational. Conversely, forcing the model to select specific facilities to open can adversely affect the optimal results.

b.13 **Assets Constraints**, a constraint can be added and set for the group of factories, fill in *0* and *4* in **Min** and **Max**, respectively. However, this setting is not necessary for this simple experiment, because the model will automatically give the optimal network structure with the lowest cost. However, to limit the number of factories opened to one or two, the corresponding option must be utilized to set a constraint.

c. Run the NO experiment.
First, conduct the data validation before running the experiment by clicking the valid button VALID to the right of the scenario name. If there is no error notification, next, set experiment parameters by enabling the setting icon, before running this numerical experiment, **Select demand variation type** is set to *Exact demand*, which means all the customer demands must be satisfied; **Select search type for N best solutions** is set to *Find N best*, and **Number of best solutions** to find is set to *1,* this means the optimal result needs to be returned in the experiment. If,

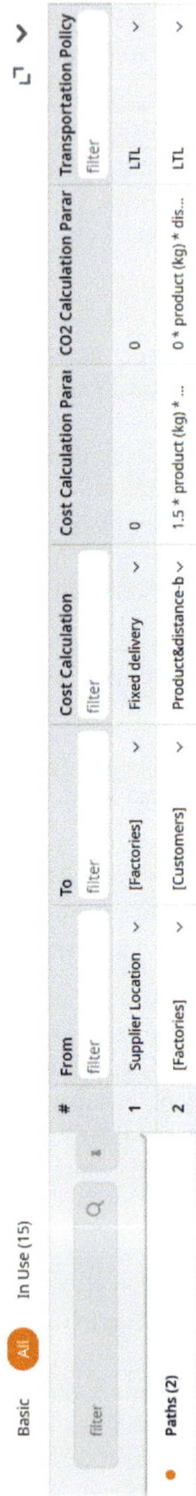

Fig. 2.5 Setting the path parameters.

for instance, *3* is put into **Number of best solutions**, anyLogistix will provide 3 optimal or near optimal solutions in the result; **Optimization time limit** is *600* seconds; For setting the **Number of threads to use**, the default value may be retained, or an appropriate value ranging from *1* to *11* can be selected based on the computer's RAM capacity. The chosen number of threads will affect the model's running time during the calculation of results. Moreover, **Finance's stats unit** is chosen *NOK* from the drop-down list; **Product stats unit** is set to *kg;* **Distance stats unit** is set to *km.*

Once all the experiment setup steps have been completed, the next step is to click the **Network optimization** in the **Experiments** panel and hit the **Run** button. When the NO experiment is completed, a new result item is created in the NO experiment branch. By clicking the **Result**, it will show the result table. This optimal result table as shown in Table 2.2 is automatically generated by anyLogistix when the NO experiment run is completed.

Table 2.2 Optimal results

#	*Description*	*Profit (NetOpt)*	*Total Costs*	*Total Demand Fulfilment, %*
1	Iteration 1: Oslo, Drammen, Bergen	–36845536	36845536	100

As can be seen from the results, to fully satisfy the customer demands, three facilities should be open, which are factory Oslo, factory Drammen, and factory Bergen. Table 2.3 presents the decisions of facility selection and demand allocation in the optimal solution. The demand from Oslo is distributed between factory Oslo and factory Drammen, while demands from other cities are met by individual factories. The capacity utilization at plants Oslo, Bergen, and Drammen are 100%, 56.9%, and 51.2%, respectively.

Table 2.3 Optimal results of facility selection and demand allocation from anyLogistix

Candidate	*Facility selection*	*Demand allocation*					
		Oslo	*Bergen*	*Trondheim*	*Stavanger*	*Drammen*	*Frederiksted*
Oslo	Open	32258		10741			
Bergen	Open		15591		8877		
Trondheim (near)	Close						
Drammen	Open	10081				6622	5311

Simultaneously, anyLogistix offers the capability to easily switch between structural and map views by selecting the view icon. In addition, access to the data dashboard and the ability to define key performance indicators (KPIs) as needed facilitate further analysis, providing deeper insights into the solutions. Refer to Figure 2.6 for visual representations.

2.4.4 Uncapacitated Network Design Problems in ALX

The uncapacitated network design problem has been a fundamental topic in operations research since the mid-20th century (Daskin et al., 1993). Unlike the capacitated network design problem, where the production capacity of factories is predetermined, a typical uncapacitated network design problem treats factory capacity as a decision variable rather than a fixed input. This approach allows the optimization model to determine the optimal scale of factories to be installed (Contreras and Fernández, 2012).

In anyLogistix, to address the previously defined problem, only some settings need to be modified as follows:

In **Product flows**, two product flows have been set up: (1) *Supplier → [Factories]*, (2) *[Factories] → [Customers]*. In the case when the capacity constraints of each factory are relaxed, the **Max Throughput** for both product flows needs to be set to *0*, which means the capacity of flow is unlimited. By default, the **Expand Sources/Destinations** setting is enabled, which indicates that the capacity of each facility is unlimited for each product.

In **Product Storage**, the **Max stock** for all factories are set to *0*. By default settings, *0* in the stock setting means the stock capacity is unlimited.

All the other modeling steps are the same as the capacitated network design problem. Then, running this model, the optimal result can be obtained with a total cost of 30749883NOK, as shown in Table 2.4, all the demands are 100% fulfilled. The optimal facility selection result indicates that two factories located in Oslo and Bergen, with estimated capacities of 65015 and 24469 units respectively, are operational to satisfy all customer demands. As shown in Table 2.5, the demands from different regions are fully met by these two factories.

Table 2.4 Optimal results of the uncapacitated network design problem

#	*Description*	*Profit (NetOpt)*	*Total Costs*	*Total Demand Fulfilment, %*
1	Iteration 1: Oslo, Bergen	–30749883	30749883	100

Table 2.5 Facility selection and demand allocation

Candidate	*Facility selection*	*Demand allocation*					
		Oslo	*Bergen*	*Trondheim*	*Stavanger*	*Drammen*	*Frederiksted*
Oslo	Open	42340		10741		6622	5311
Bergen	Open		15591		8877		
Trondheim (near)	Close						
Drammen	Close						

When comparing the results of facility selection and demand allocation from anyLogistix with those obtained using mathematical optimization models

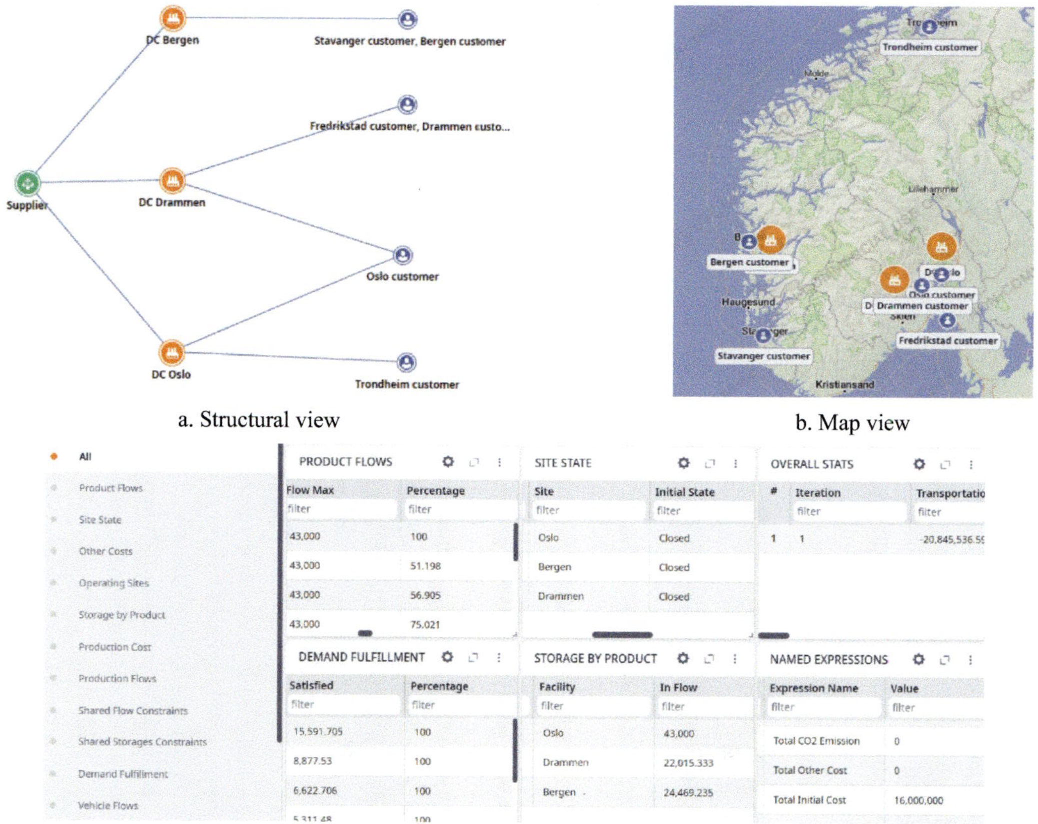

a. Structural view

b. Map view

c. KPIs data dashboard

Fig. 2.6 Data visualization and analysis in anyLogistix.

in Chapter 1, the optimal decisions align consistently, though the objective values for total cost differ slightly. Specifically, the total costs from the gravity location model and the uncapacitated network design model are 36385930 NOK and 30832893 NOK, respectively, compared to 36845536 NOK and 30749883 NOK from the GFA model and NO model. This variation is primarily attributed to the use of different GIS maps to obtain the distance matrix, affecting the calculation of distance-based transportation costs.

Comparing results across both chapters, the digital logistics twin has proven effective for distribution network design, significantly reducing modeling and coding time. Moreover, through seamless data conversion, the optimal distribution network can be further evaluated using dynamic simulation in a more comprehensive environment and operational policies.

2.5 Conclusion

In this chapter, we explored the transformative potential of digital twin technology within the context of Industry 5.0, particularly focusing on its application in distribution and logistics network design. The integration of digital twins offers a robust framework for enhancing decision-making processes and operations by accurately mirroring the physical logistics processes and operations in real-time. This capability is achieved through the seamless integration of advanced technologies, i.e., IoT, CPS, AI, big data analytics, optimization, and simulation, which collectively form a comprehensive platform for improved sustainability, responsiveness, and digitization. By providing a dynamic and self-evolving virtual model of real-world systems, digital twins enable organizations to make informed, proactive, and reactive decisions. This not only enhances operational efficiency in supply chain and logistics management but also supports the broader goals of Industry 5.0, emphasizing human-centric and sustainable business practices.

This chapter provides a comprehensive overview of the theoretical and methodological development of digital twins, with a focus on their applications in logistics and supply chain management. Numerical experiments are provided to show how digital logistics twins can be built to solve distribution network design problems using anyLogistix. By utilizing a common and shared data model, GFA, NO, and dynamic simulation can be seamlessly integrated into anyLogistix to enhance decision-making for logistics network configurations. The results, compared with those from the optimization models in Chapter 1, demonstrate the effectiveness of digital logistics twins in solving distribution network design problems.

On one hand, by integrating analytical optimization models for different location and network design problems, digital logistics twins significantly reduce modeling and coding time, thereby accelerating decision-making processes. Furthermore, distribution system designs are better visualized and can be further evaluated by dynamic simulation across comprehensive scenarios, configuration updates, and operational conditions (Sun et al., 2024). However, on the other hand, a notable limitation of digital logistics twins is the lack of flexibility in model configurations. As an off-the-shelf solution, accommodating new conditions or constraints can be challenging. For example, when implementing a single-source policy or adopting electric trucks in a logistics system, the digital logistics twin may be unable to accommodate the new requirements or identify optimal locations for charging stations due to the lack of built-in models.

Declaration of Generative AI and AI-assisted Technology in the Writing Process

During the preparation of the manuscript, the authors used ChatUiT (a tailored version of ChatGPT) and ChatGPT o1-mini to polish the scientific writing. After using these tools, the authors reviewed and edited the content as needed and take full responsibility for the content of the paper.

Acknowledgment

The research is supported by the Industry 5.0 enabled Smart Logistics: A Global Perspective project funded by the Norwegian Directorate for Higher Education and Skills (HK-dir) under the Utforsk Programme (UTF-2021/10166), the UiT Aurora project MASCOT, and the International Cooperation Plan under the Department of Science and Technology in Sichuan Province (No.2024YFHZ0315).

References

Ailawadi, S.C. and Singh, P.R. 2011. *Logistics management*, PHI Learning Pvt. Ltd.

Alx 2024. anyLogistix: software for supply chain design and optimization. Available from: https://www.anylogistix.com/ [accessed Nov 15 2024].

Barata, J. and Kayser, I. 2024. How will the digital twin shape the future of industry 5.0? *Technovation*, 134, 103025.

Bhandal, R., Meriton, R., Kavanagh, R.E. and Brown, A. 2022. The application of digital twin technology in operations and supply chain management: a bibliometric review. *Supply Chain Management: An International Journal*, 27, 182–206.

Butt, J. 2020. A Strategic Roadmap for the Manufacturing Industry to Implement Industry 4.0. *Designs*, 4, 11.

Chen, Z., Surendraarcharyagie, K., Granland, K., Chen, C., Xu, X., Xiong, Y., Davies, C. and Tang, Y. 2024. Service oriented digital twin for additive manufacturing process. *Journal of Manufacturing Systems*, 74, 762–776.

Contreras, I. and Fernández, E. 2012. General network design: A unified view of combined location and network design problems. *European Journal of Operational Research*, 219, 680–697.

Daskin, M., Hurter, A. and Van Buer, M. 1993. Toward an integrated model of facility location and transportation network design, The Transportation Center. Northwestern University, Working Paper. Evanston, IL, USA.

Farahani, R.Z., Rezapour, S., Drezner, T. and Fallah, S. 2014. Competitive supply chain network design: An overview of classifications, models, solution techniques and applications. *Omega*, 45, 92–118.

Figueiras, P., Lourenço, L., Costa, R., Graça, D., Garcia, G. and Jardim-Gonçalves, R. Big data provision for digital twins in industry 4.0 logistics processes. 2021 IEEE International Workshop on Metrology for Industry 4.0 & IoT (MetroInd4. 0&IoT), 2021. Ieee, 516–521.

Fuller, A., Fan, Z., Day, C. and Barlow, C. 2020. Digital twin: Enabling technologies, challenges and open research. *IEEE access*, 8, 108952–108971.

Gai, K., Zhang, Y., Qiu, M. and Thuraisingham, B. 2022. Blockchain-enabled service optimizations in supply chain digital twin. *IEEE Transactions on Services Computing*, 16, 1673–1685.

Ghaderi, A. and Jabalameli, M.S. 2013. Modeling the budget-constrained dynamic uncapacitated facility location–network design problem and solving it via two efficient heuristics: A case study of health care. *Mathematical and Computer Modelling*, 57, 382–400.

Grieves, M. and Vickers, J. 2017. Digital twin: Mitigating unpredictable, undesirable emergent behavior in complex systems. *Transdisciplinary perspectives on complex systems: New findings and approaches*, 85–113.

Ivanov, D. 2023. Intelligent digital twin (iDT) for supply chain stress-testing, resilience, and viability. *International Journal of Production Economics*, 263, 108938.

Ivanov, D. 2024. Conceptualisation of a 7-element digital twin framework in supply chain and operations management. *International Journal of Production Research*, 62, 2220–2232.

Ivanov, D. and Dolgui, A. 2021. A digital supply chain twin for managing the disruption risks and resilience in the era of Industry 4.0. *Production Planning & Control*, 32, 775–788.

Ivanov, D., Dolgui, A., Das, A. and Sokolov, B. 2019. Digital supply chain twins: Managing the ripple effect, resilience, and disruption risks by data-driven optimization, simulation, and visibility. *Handbook of ripple effects in the supply chain*, 309–332.

Jacoby, M. and Usländer, T. 2020. Digital twin and internet of things—Current standards landscape. *Applied Sciences*, 10, 6519.

Jafari, N., Azarian, M. and Yu, H. 2022. Moving from Industry 4.0 to Industry 5.0: what are the implications for smart logistics? *Logistics*, 6, 26.

Jesus, V., Kalaitzi, D., Batista, L. and Lopez, N.L. 2024. Digital Twins of Supply Chains: A Systems Approach. *IEEE Transactions on Engineering Management*, 71, 14915–14932.

Kamble, S.S., Gunasekaran, A., Ghadge, A. and Raut, R. 2020. A performance measurement system for industry 4.0 enabled smart manufacturing system in SMMEs-A review and empirical investigation. *International journal of production economics*, 229, 107853.

Kamble, S.S., Gunasekaran, A., Parekh, H., Mani, V., Belhadi, A. and Sharma, R. 2022. Digital twin for sustainable manufacturing supply chains: Current trends, future perspectives, and an implementation framework. *Technological Forecasting and Social Change*, 176, 121448.

Kane, T.M. 2012. *Military logistics and strategic performance*, Routledge.

Kobayashi, T., Hata, H., Fukae, K., Sato, H., Tanimoto, S. and Kanai, A. Digital Twin Configuration Method for Public Services by Citizens. 2023 IEEE International Conference on Consumer Electronics (ICCE), 2023. IEEE, 01–03.

Leng, J., Zhang, H., Yan, D., Liu, Q., Chen, X. and Zhang, D. 2019. Digital twin-driven manufacturing cyber-physical system for parallel controlling of smart workshop. *Journal of ambient intelligence and humanized computing*, 10, 1155–1166.

Li, L., Aslam, S., Wileman, A. and Perinpanayagam, S. 2021. Digital twin in aerospace industry: A gentle introduction. *IEEE Access*, 10, 9543–9562.

Liu, J., Cao, X., Zhou, H., Li, L., Liu, X., Zhao, P. and Dong, J. 2021. A digital twin-driven approach towards traceability and dynamic control for processing quality. *Advanced Engineering Informatics*, 50, 101395.

Liu, S., Lu, Y., Li, J., Shen, X., Sun, X. and Bao, J. 2023. A blockchain-based interactive approach between digital twin-based manufacturing systems. *Computers & Industrial Engineering*, 175, 108827.

Lummus, R.R., Krumwiede, D.W. and Vokurka, R.J. 2001. The relationship of logistics to supply chain management: developing a common industry definition. *Industrial management & data systems*, 101, 426–432.

Lv, Z. 2023. Digital twins in industry 5.0. *Research*, 6, 0071.

Maheshwari, P., Kamble, S., Belhadi, A., Venkatesh, M. and Abedin, M.Z. 2023. Digital twin-driven real-time planning, monitoring, and controlling in food supply chains. *Technological Forecasting and Social Change*, 195, 122799.

Mandolla, C., Petruzzelli, A. M., Percoco, G. and Urbinati, A. 2019. Building a digital twin for additive manufacturing through the exploitation of blockchain: A case analysis of the aircraft industry. *Computers in industry*, 109, 134–152.

Mo, D.-H., Tien, C.-L., Yeh, Y.-L., Guo, Y.-R., Lin, C.-S., Chen, C.-C. and Chang, C.-M. 2023. Design of digital-twin human-machine interface sensor with intelligent finger gesture recognition. *Sensors*, 23, 3509.

Moshood, T.D., Nawanir, G., Sorooshian, S. and Okfalisa, O. 2021. Digital twins driven supply chain visibility within logistics: A new paradigm for future logistics. *Applied System Innovation*, 4, 29.

Park, K. T., Lee, J., Kim, H.-J. and Noh, S. D. 2020. Digital twin-based cyber physical production system architectural framework for personalized production. *The International Journal of Advanced Manufacturing Technology*, 106, 1787–1810.

Preut, A., Kopka, J.-P. and Clausen, U. 2021. Digital Twins for the Circular Economy. *Sustainability*, 13, 10467.

Qi, Q. and Tao, F. 2018. Digital twin and big data towards smart manufacturing and industry 4.0: 360 degree comparison. *Ieee Access*, 6, 3585–3593.

Qi, Q., Tao, F., Zuo, Y. and Zhao, D. 2018. Digital twin service towards smart manufacturing. *Procedia Cirp*, 72, 237-242.

Rabah, S., Zacharewicz, G. and Chapurlat, V. 2022. Digital Twin for services (DT4S): Conceptual strategy. *IFAC-PapersOnLine*, 55, 3256–3261.

Rigó, L., Fabianová, J., Lokšík, M. and Mikušová, N. 2024. Utilising Digital Twins to Bolster the Sustainability of Logistics Processes in Industry 4.0. *Sustainability*, 16, 2575.

Rushton, A., Croucher, P. and Baker, P. 2022. *The handbook of logistics and distribution management: Understanding the supply chain*, Kogan Page Publishers.

Schenkel, M., Caniëls, M.C., Krikke, H. and Van Der Laan, E. 2015. Understanding value creation in closed loop supply chains–Past findings and future directions. *Journal of Manufacturing Systems*, 37, 729–745.

Schleich, B., Anwer, N., Mathieu, L. and Wartzack, S. 2017. Shaping the digital twin for design and production engineering. *CIRP annals*, 66, 141–144.

Serrano, A., Kalenatic, D., López, C. and Montoya-Torres, J.R. 2023. Evolution of Military Logistics. *Logistics*, 7, 22.

Seuring, S. and Gold, S. 2013. Sustainability management beyond corporate boundaries: from stakeholders to performance. *Journal of cleaner production*, 56, 1–6.

Shen, T. and Li, B. 2024. Digital twins in additive manufacturing: a state-of-the-art review. *The International Journal of Advanced Manufacturing Technology*, 131, 63–92.

Singh, M., Fuenmayor, E., Hinchy, E.P., Qiao, Y., Murray, N. and Devine, D. 2021. Digital twin: Origin to future. *Applied System Innovation*, 4, 36.

Stark, R., Fresemann, C. and Lindow, K. 2019. Development and operation of Digital Twins for technical systems and services. *CIRP Annals*, 68, 129–132.

Sun, X., Yu, H. and Solvang, W.D. Industry 4.0 and sustainable supply chain management. Advanced Manufacturing and Automation X 10, 2021. Springer, 595–604.

Sun, X., Yu, H. and Solvang, W. D. System Integration for Smart Reverse Logistics Management. 2022 IEEE/SICE International Symposium on System Integration (SII), 2022. IEEE, 821–826.

Sun, X., Yu, H. and Solvang, W. D. A Digital Reverse Logistics Twin for Improving Sustainability in Industry 5.0. IFIP International Conference on Advances in Production Management Systems, 2023a. Springer, 273–286.

Sun, X., Yu, H., Solvang, W.D. and Govindan, K., 2024. A two level decision support framework for reverse logistics network design considering technology transformation in Industry 4.0: a case study in Norway. *The International Journal of Advanced Manufacturing Technology*, 134(1), 389–413.

Tao, F., Cheng, J., Qi, Q., Zhang, M., Zhang, H. and Sui, F. 2018. Digital twin-driven product design, manufacturing and service with big data. *The International Journal of Advanced Manufacturing Technology*, 94, 3563–3576.

Van Der Valk, H., Strobel, G., Winkelmann, S., Hunker, J. and Tomczyk, M. 2022. Supply chains in the era of digital twins–a review. *Procedia Computer Science*, 204, 156–163.

Wang, H., Lv, L., Li, X., Li, H., Leng, J., Zhang, Y., Thomson, V., Liu, G., Wen, X. and Sun, C. 2023. A safety management approach for Industry 5.0′ s human-centered manufacturing based on digital twin. *Journal of Manufacturing Systems*, 66, 1–12.

Wilhelm, J., Beinke, T. and Freitag, M. Improving Human-Machine Interaction with a Digital Twin: Adaptive Automation in Container Unloading. Dynamics in Logistics: Proceedings of the 7th International Conference LDIC 2020, Bremen, Germany, 2020. Springer, 527–540.

Yu, H. and Sun, X. 2024. Uncertain remanufacturing reverse logistics network design in industry 5.0: Opportunities and challenges of digitalization. *Engineering Applications of Artificial Intelligence*, 133, 108578.

3 Artificial Intelligence and Data-driven Smart Transportation and Logistics: Unsupervised Learning Models and Ensemble Learning Models

Tao Peng[1], Lin Ma[2], Xinchi Dong[3], Zihan Zhang[4], Yuxin He[5], Mi Gan[6*] and Hao Yu[7*]

3.1 Introduction

With the rapid advancement of information technology and the exponential growth in data generation, the application of big data has become increasingly pervasive across various sectors, particularly in logistics and transportation. These industries produce vast quantities of data on a daily basis, capturing critical aspects such as vehicle positioning, traffic conditions, user demand, and operational performance metrics. The effective utilization of this extensive data can not only enhance the efficiency of current transportation and logistics processes but also serve as an essential resource for informed decision-making, potentially drastically reshaping the future of the industry. Logistics and transportation systems, however, are characterized by their inherent complexity, involving dynamic data that continuously evolves over time. This complexity poses significant challenges for traditional analytical methodologies, which often struggle to address the diverse and fluctuating nature of the data involved (Li, 2024). In response to these challenges, the concept of data-driven smart logistics and transportation has emerged as a highly focused research area, particularly in the context of recent advancements in big data analytics, artificial intelligence (AI), and machine learning technologies. Research in this domain is dedicated to harnessing the power of extensive datasets to optimize the performance of transportation and logistics systems (Zhu, 2018). Current research predominantly concentrates on several key areas: Prediction of routes, demand, and delivery time, optimization of warehouse layout and delivery routes, construction of complex networks to facilitate better logistics planning, and exploration of transportation decarbonization strategies to promote sustainability. Noteworthy progress has been made in these areas. For instance, advanced analytical techniques that leverage historical traffic data, real-time road condition

[1-6] School of Transportation and Logistics, Southwest Jiaotong University, Chengdu, China.
[7] Department of Industrial Engineering, UiT—The Arctic University of Norway, 8514 Narvik, Norway.
* Corresponding author: migan@swjtu.edu.cn, hao.yu@uit.no

DOI: 10.1201/9781003466451-3

updates, integrated with machine learning algorithms enable logistics companies to anticipate traffic congestion in specific timeframes and geographic locations. This proactive approach facilitates the preemptive adjustment of delivery routes, effectively circumventing congested zones. By implementing these strategies, companies can devise optimal delivery plans for each shipment, significantly minimizing both transportation duration and operational costs (Osaba et al., 2020; Tsang et al., 2018). Moreover, through the analysis of historical order patterns, seasonal demand variability, and overarching market trends, businesses can enhance their demand forecasting capabilities. This increased precision in demand forecasting allows for proactive inventory management, resulting in optimized inventory levels and reduced waste and overstock. As a consequence, organizations can achieve higher warehouse utilization rates, ultimately contributing to a more agile and responsive supply chain (Kumar et al., 2024).

To achieve these goals, a variety of advanced technical approaches have been adopted, with unsupervised learning models, ensemble learning models, and artificial neural networks being the predominant methods. These technologies not only demonstrate strong capabilities in data processing and analysis but also provide new solutions for optimizing logistics and transportation systems. In the following sections, we will delve into the methodology of these models and their applications in the field of smart logistics and transportation. Specifically, this chapter focuses on unsupervised learning methods and ensemble learning methods, and Chapter 4 introduces the application of artificial neural networks in smart transportation and logistics.

The chapter is organized as follows. Section 3.2 provides a literature review on data-driven smart transportation and logistics. Sections 3.3 and 3.4 introduce unsupervised machine learning methods and ensemble learning methods, respectively. Their applications in smart transportation and logistics are also discussed in these sections. Both sections also provide details of one machine learning method and two ensemble learning methods. Section 3.5 concludes the chapter.

3.2 Data-driven Smart in Transportation and Logistics

In the realm of predicting demand and delivery times using big data combined with AI, Chen et al. (2023) proposed a novel deep learning method known as the Bidirectional Distance Isotropic Gated Recurrent Unit (BDIGRU). This method is integrated into the deep learning framework of neural networks for travel time prediction analysis and route planning. It learns high-level features directly from large traffic datasets and employs a time-sequenced attention mechanism for reconstruction, facilitating an end-to-end recursive learning process. The proposed method is utilized to analyze and predict random travel times under various traffic conditions, particularly during congested scenarios, in order to determine the optimal vehicle routes that minimize travel time under uncertainties. Yang et al. (2023) developed a deep neural network based on a deep gravity framework to predict intercity heavy truck traffic flows, presenting an advanced model for human mobility forecasting. This study leverages extensive data related to heavy truck movements, freight locations, road networks, and land use to train the model, and its performance is validated through comparisons with traditional gravity models. The findings have significant implications for urban planning, traffic management, and logistics operations. In the context of site selection and travel pattern exploration using big data, Zheng et al. (2024) developed a deep learning-based method for locating high-speed rail freight stations. This method aims to uncover the potential site selection patterns for high-speed rail freight stations based on existing high-speed rail express logistics base construction projects. The research initially constructs site selection conditions across three dimensions: market demand, locational transportation, and developmental environment, along with ten corresponding indicators. A site selection model is developed based on fundamental features and deep learning algorithms, employing genetic algorithms to optimize hyperparameters. The site selection results further validate that the deep learning model established in this study effectively learns the actual site selection patterns for high-speed rail freight stations, facilitating a more scientific and structured decision-making process for site selection. He et al. (2024) proposed a framework for constructing and classifying truck travel chains based on trajectory data in the context of intercity freight transportation. From the perspective of travel chains, this framework explores the travel behaviors and characteristics of trucks in intercity freight through both numerical and route features. It aims to provide a deep understanding and analysis of truck drivers' travel preferences, thereby offering effective suggestions for the efficient management of intercity freight. In addition, big data and AI technologies are frequently employed in research focused on transportation decarbonization and sustainability. Yao et al. (2023) analyzed a large volume of trajectory data to decompose the relationship between travel and emissions from the perspective of travel chains. They introduced a method for calculating emissions from road freight vehicles that considers variations in loading states based on different travel purposes. Comparisons with previous studies indicate that the proposed method for estimating truck emissions demonstrates high accuracy and reliability when accounting for various trip purposes. Peng et al. (2024) utilized a substantial dataset of truck GPS data from Chengdu, employing the International Vehicle Emissions (IVE) model to estimate freight-related emissions. Furthermore, they conducted a detailed analysis using Gradient Boosting Regression Trees (GBRT) to explore the impact of urban built environment factors, i.e., land use, traffic design, and density, on these emissions. This research provides new perspectives and methodologies for managing transportation-related carbon emissions. Utilizing big data to construct freight maps and complex networks is also an important area of research. Bai et al. (2023) employed high-frequency vessel movement data to develop a Global Liner Shipping Network (GLSN). Community Penetration Methods (CPMs) were then used to identify overlapping community structures and key nodes within the network. Following this, they analyzed network resilience from both static and dynamic perspectives. Building on the existing techniques for map construction, Mao et al. (2023) addressed the limitations of current electronic map construction methods in the bulk freight sector by proposing a data-driven framework for logistics map construction utilizing multi-source logistics data. The findings provide substantial practical insights for technological innovations in logistics map construction and offer new solutions for enhancing cost efficiency in bulk logistics, thus possessing significant theoretical importance and practical value.

The field of data-driven smart logistics and transportation has garnered significant attention in recent years, leveraging big data analytics and AI technologies to enhance operational efficiency, optimize decision-making processes, and improve service quality. Current research primarily focuses on several key areas, including demand forecasting, route optimization, traffic management, and sustainability initiatives. By employing advanced techniques such as machine learning and deep learning, researchers can effectively analyze the vast and dynamic datasets generated by transportation systems, enabling more accurate demand predictions and efficient route planning. Additionally, there is a growing emphasis on low-carbon transportation solutions aimed at minimizing carbon emissions associated with logistics activities (Ben Ayed et al., 2015). Despite the advancements made in this field, several challenges remain. First, issues related to data privacy and security are pressing, as the extensive collection and processing of large datasets raise concerns about user privacy and data protection (Shozi and Mtsweni, 2016; Wang et al., 2019). Second, the lack of standardization in data formats and models hinders interoperability among different systems, which can negatively impact overall efficiency (Al-Sai et al., 2019). Furthermore, the scalability of data-driven solutions in diverse logistics environments poses a significant challenge that requires further investigation (Li, 2024). Future research directions could focus on the integration of the Internet of Things (IoT) with big data analytics to enhance data collection and real-time monitoring capabilities (Zafarzadeh et al., 2021). Additionally, exploring strategies to bolster the resilience and adaptability of supply chains in the face of disruptions is critical (Bai et al., 2023). The development of personalized logistics services through customer data analytics presents another promising avenue for research (Chung, 2021). Lastly, examining the impact of regulatory frameworks on the adoption of smart logistics technologies is an important topic that warrants deeper exploration. Addressing these challenges and pursuing these research directions not only holds theoretical significance but also provides valuable insights and solutions for the practical advancement of smart logistics and transportation systems.

3.3 Unsupervised Machine Learning Methods and Their Applications

3.3.1 Unsupervised Machine Learning Methods

Unsupervised learning is a branch of machine learning that can process unlabeled datasets and discover the internal structure and patterns within the data. Its core lies in exploring the intrinsic representation and latent rules of the data, thereby achieving an understanding and modeling of the data.

The objectives of unsupervised learning can include, but are not limited to: exploring the distribution of the data to understand its composition; uncovering hidden patterns or regularities within the data; reducing the dimensionality of the data to make it easier to handle and visualize; and identifying natural groupings or clusters within the data. Unsupervised learning algorithms typically aim to maximize certain intrinsic properties of the data, i.e., similarity, compactness, or statistical consistency, to reveal important features and relationships within the dataset.

Due to its flexibility and broad applicability, unsupervised learning methods have been applied in many fields, including transportation and logistics. For example, Gong et al. (2018) applied the Density-Based Spatial Clustering of Applications with Noise (DBSCAN) algorithm from unsupervised learning to identify stop points during the transport routes of trucks from a large amount of trajectory data. Duan et al. (2020) utilized the K-means algorithm to categorize multi-day trip chains of trucks, analyzing the matching rules between travel chain patterns and truck types. Deng et al. (2023) used Gaussian Mixture Model (GMM) to classify passenger vehicles, investigating differences in travel time and route choice among various types of vehicles. Table 3.1 summarizes the advantages, disadvantages, and applications of the aforementioned methods.

For the practical application introduced in this section, the GMM model is used to classify the trip chains of trucks, thus revealing the characteristics and differences of various categories of truck trip chains.

Table 3.1 Overview of unsupervised machine learning methods

Algorithms	*Pros and cons*	*Applications*
K-means clustering algorithm (K-means)	*Pros*: Simple and efficient, suitable for large data sets; Can quickly divide data into a pre-specified number of clusters. *Cons*: The number of clusters needs to be pre-specified and easy to fall into the local optimal solution; High requirements on the form of data distribution.	K-means is widely used for vehicle stop point classification and travel time prediction. (Aziz et al., 2016; Wang et al., 2016)
K-means ++ clustering algorithm (K-means++)	*Pros*: Smarter selection of initial clustering centers and reduce the risk of falling into local optimal solutions. *Cons*: Slightly increase the computational complexity. Still need to pre-specify the number of clusters.	K-means++ is widely used in vehicle travel feature classification (Duan et al., 2020)

Contd.

Algorithms	*Pros and cons*	*Applications*
Density-Based Spatial Clustering of Applications with Noise (DBSCAN)	*Pros*: Can automatically find clusters of any shape without specifying the number of clusters in advance. Ability to identify and eliminate noise points. *Cons*: High requirements for the selection of clustering parameters; Maybe not efficiently cluster Data sets with large density variations.	DBSCAN is widely used to identify vehicle stop points. (Gong et al., 2018a)
Gaussian Mixture Model (GMM)	*Pros*: Can flexibly model complex data distributions and allow each data point to belong to multiple clusters with probabilistic representation. *Cons*: Sensitive to initial values; High computational complexity; Require pre-specifying the number of clusters in advance.	GMM is widely used in vehicle travel mode classification. (Deng et al., 2023)

3.3.2 Gaussian Mixture Model (GMM)

a. Methodology

GMM is an approach within the realm of unsupervised machine learning, which aims at finding a mix of multivariate Gaussian probability distributions to better fit the data (Deng et al., 2023). A key advantage of GMM is its ability to capture complex distribution shapes since it allows clusters to have different shapes and sizes, rather than being limited to spherically symmetric distributions. Furthermore, GMM provides a probabilistic cluster assignment mechanism, meaning it can offer soft estimates of the probability that data points belong to different clusters, rather than making hard assignments. These features make GMM particularly effective when dealing with datasets that contain overlapping or non-convex-shaped clusters. The formula are given in Eqs. (3.1-3.4):

$$P(Xi \mid \theta) = \sum_{k=1}^{k} a_k p_k (X_i \mid u_k, \Sigma_k) \qquad \text{...(3.1)}$$

$$\theta = (a_1, a_2, \ldots a_K, u_1, u_2, u_k, \ldots, \Sigma_1, \Sigma_2, \ldots, \Sigma_K) \qquad \text{...(3.2)}$$

$$\sum_{k=1}^{K} a_k = 1, 0, < a_k < 1 \qquad \text{...(3.3)}$$

$$p_k(X_i \mid u_k, \Sigma_k) = \frac{1}{(2\pi)^{\frac{n}{2}} |\Sigma_i|^{\frac{1}{2}}} \exp\left(-\frac{(X_i - u_k)^T \Sigma_i^{-1} (X_i - u_k)}{2} \right) \qquad \text{...(3.4)}$$

Where $X_i = [x_1, x_2, \dots x_n]$ represents an observation data with n dimensional features from a total of N samples; K is the number of single Gaussian distributions in GMM; θ represents the set of parameters to be estimated in the model; a_k represents the probability that X_i belongs to the kth Gaussian distribution. u_k and Σ_k represent the mean and covariance matrix of the kth Gaussian distribution; $p_k(X_i | u_k, \Sigma_k)$ is the probability density function of the kth Gaussian distribution.

GMM requires estimation of parameters a_k, u_k, and Σ_k. Typically, the Expectation-Maximization (EM) algorithm is used for optimization, and the procedures of the method are as follows:

(1) Initialize the parameters.

(2) E-step: Based on the current parameters, calculate the posterior probability γ_{ik} of each sample X_i belonging to the kth Gaussian distribution (Eq. (3.5)).

$$\gamma_{ik} = \frac{a_k p_k(X_i | u_k, \Sigma_k)}{\sum_{k=1}^{K} a_k p_k(X_i | u_k, \Sigma_k)} \quad \text{...(3.5)}$$

(3) M-step: Maximize the log-likelihood function to compute new model parameters and update the parameters for the next iteration (Eqs. (3.6-3.9)).

$$L(\theta | X_1, \dots X_N) = \sum_{i=1}^{N} \ln \left\{ \sum_{k=1}^{K} a_k p_k(X_i | u_k, \Sigma_k) \right\} \quad \text{...(3.6)}$$

$$u_k = \frac{\sum_{i=1}^{N} (\gamma_{ik} X_i)}{\sum_{i=1}^{N} \gamma_{ik}} \quad \text{...(3.7)}$$

$$\Sigma_k = \frac{\sum_{i=1}^{N} \gamma_{ik} (X_i - u_k)(X_i - u_l)^T}{\sum_{i=1}^{N} \gamma_{ik}} \quad \text{...(3.8)}$$

$$a_k = \frac{\sum_{i=1}^{N} \gamma_{ik}}{N} \quad \text{...(3.9)}$$

(4) Repeat the E-step and M-step iteratively, continuously updating the parameters in GMM until convergence is achieved.

b. Practical Application in Logistics

b.1 Data description

We utilized GMM clustering to classify trip chains extracted from a nationwide set of freight truck trajectories in China, aiming to analyze the travel characteristics of truck drivers. The trajectory dataset of freight trucks used in our case is obtained from an OFEX platform based in China. It includes a total of 13,356 trip chains from 2000 trucks over a period between October 2018 and November 2018.

b.2 Input indicators

Four indicators of travel chains were selected as the input data for GMM, which are trip chain distance, number of destination cities, trip chain duration, and average dwell time in destination cities. The distribution of these four indicators is shown in Figure 3.1.

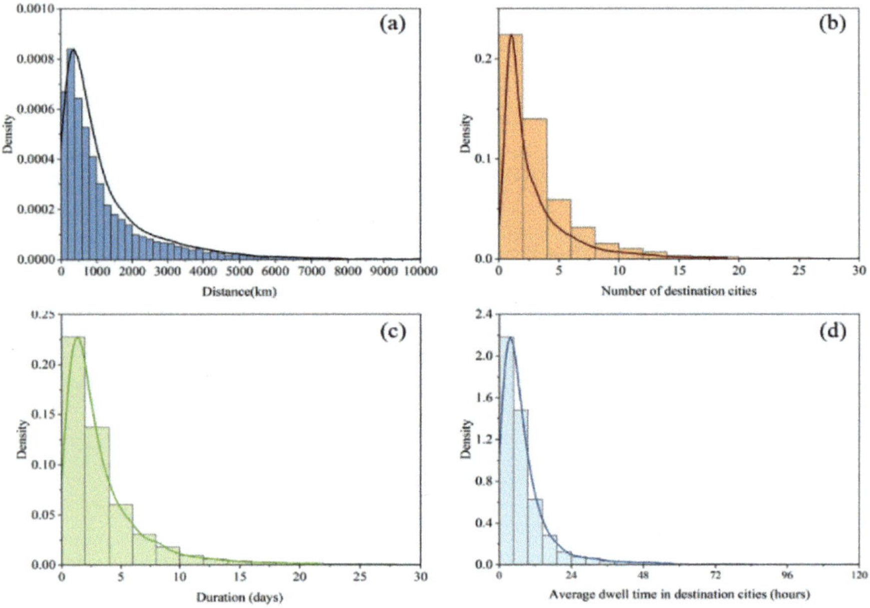

Fig. 3.1 The distribution of trip chain indicators.

b.3 Model evaluation and effect

To determine the optimal number of Gaussian distributions in GMM, the Bayesian Information Criterion (BIC) can be used. BIC ensures that the model has better fitting capabilities while also maintaining its generalization ability. The formula is is given in Eq. (3.10):

$$\mathrm{BIC} = -2\ln(L) + K\ln(N) \qquad ...(3.10)$$

BIC represents the value of the Bayesian Information Criterion; L represents the value of the likelihood function; K represents the number of Gaussian distributions; N represents the sample size. A lower BIC indicates better model performance. When $K = 6$, the BIC reaches the minimum value of our case.

In addition, we used the CH index to compare the clustering performance of GMM with several other clustering algorithms at their optimal number of clusters. A higher CH index indicates better clustering performance. As shown in Table 3.2, GMM has the highest CH index, indicating that under this k value, GMM performs very well in the classification of travel chains.

Table 3.2 Comparison of performance of various unsupervised learning methods

Algorithms	*GMM*	*K-means*	*K-means++*	*PAM*
CH	7021	5889	6037	4876

b.4 Result and analysis

As a result, we classified the trip chains into six categories, with each category representing a specific travel chain pattern. Table 3.3 presents the indicator characteristics of these 6 trip chain patterns. Of the total trip chains, 40.6% of them belong to Category 4, which has the shortest average travel distance and only involves one destination city. This pattern is likely a one-way travel back and forth within the same province or neighboring provinces, which is the most popular case. Category 1, Category 2, and Category 3 have similar proportions in the number of trip chains, they exhibit significant differences in average distance, average number of destination cities, and average travel duration. Category 5 and Category 6 have the lowest proportion of trip chains. Category 5 has longer travel distance compared to the other categories such as Category 1 and Category 3, but it has fewer destination cities. Additionally, this category has a significantly higher average dwell time at the destination cities compared to the other five categories. Category 6 has a much longer average distance, a more average number of destination cities, and a longer average duration compared to the other five categories. Apparently, Category 5 and Category 6 are less preferred by trucks. Overall, the analysis indicates that different trip chain patterns have distinct characteristics, and some patterns are more common and preferred by trucks for intercity transportation.

Table 3.3 Six trip chain patterns

Category	*Average distance (km)*	*Average number of destination cities*	*Average travel duration (day)*	*Average dwell time at destination cities (h)*	*Proportion*
1	683.04	2.37	1.82	7.81	15.8%
2	2674.53	6.54	6.11	7.88	15.6%
3	1301.55	3.51	3.33	4.29	19.1%
4	383.93	1.00	1.35	8.15	40.6%
5	1805.71	2.30	7.33	52.74	4.0%
6	5653.36	12.81	15.15	8.38	5.0%

3.4 Ensemble Learning Methods and Their Applications

3.4.1 Ensemble Learning Methods

Ensemble learning is a powerful machine learning paradigm where multiple models (often referred to as "weak learners") are combined to solve a particular problem. The primary objective of ensemble methods is to improve the predictive performance of models by integrating the strengths of several individual models to achieve better results than any single one could on its own (Mohammed and Kora, 2023). There are several key methods of ensemble learning, including bagging, boosting, and stacking. Bagging (Bootstrap Aggregating) works by training multiple instances of the same algorithm on different subsets of the training data, created through random sampling with replacement. Each model makes predictions independently, and their outputs are combined, usually through averaging for regression tasks or majority voting for classification tasks. A popular example of a bagging method is the RF algorithm, which builds a multitude of decision trees and merges their predictions.

Boosting refers to a sequential ensemble technique where models are trained one after another, with each new model focusing on the errors made by the previous ones. The idea is to give more weight to misclassified instances, thereby improving the performance of the model on difficult cases. Boosting often results in a strong learner that can outperform individual models. Notable boosting algorithms include AdaBoost and Gradient Boosting Machines (GBM), which adjust the weights of instances based on the previous model's performance.

Stacking involves training multiple base models and then using another model, often called a meta-learner, to combine their predictions. The base models can be of different types and architectures, allowing for greater diversity. The meta-learner takes the predictions of the base models as input and learns to optimize the final output. Stacking can harness the strengths of various models to achieve better results than any single model alone.

In general, ensemble learning improves the generalization ability of a model by combining multiple learners. Boosting focuses on reducing bias, Bagging focuses on reducing variance, and Stacking focuses on the nonlinear combination of the model. In practical application, selecting the appropriate ensemble learning method can significantly improve the performance of the model. Table 3.4 summarizes the advantages, disadvantages, and applications of the common ensemble learning methods.

Table 3.4 Overview of ensemble learning methods

Algorithms	*Pros and cons*	*Applications*
RF (Random Forest)	It has good robustness and is suitable for a variety of data scenarios, but its training time is long, and it is difficult to explain the specific decision process.	An analysis of motorcycle accidents in Rawalpindi, Pakistan, aimed at determining risk factors (Mansoor et al., 2023). Identify and classify road damage (Dugalam and Prakash, 2024).
XGBoost	It is suitable for large-scale data, but the parameter adjustment is complex, and it is easy to overfit small data sets.	To investigate factors affecting injury severity at pedestrian crossing locations (Goswamy et al., 2023). Understanding user equilibrium states of road networks (Chen et al., 2024).

Contd.

Algorithms	*Pros and cons*	*Applications*
GBDT (Gradient-Boosted Decision Tree)	It is effective in nonlinear problems, feature importance evaluation, and small data sets, but it is slow in training and sensitive to parameters and data distribution.	Analysis of the relationship between the number of accidents and the traffic characteristics of the grid areas (Yang et al., 2022). Predicting Bus Passenger Flow and Prioritizing Influential Factors (Wu et al., 2021)
AdaBoost	It can deal with nonlinear problems and prevent overfitting well, but it is sensitive to noise and outliers, has a long training time, and requires high parameters and weak classifier selection.	The types of railway accidents that may occur in the future are predicted, and the main causes of railway accidents are analyzed (Meng et al., 2022). A license plate dataset of 950 real-world images was detected (Yao and Yi, 2014).

3.4.2 Random Forest (RF) Model

a. Methodology

RF Model is an ensemble learning technique used for classification and regression tasks. It combines the predictions of multiple decision trees to improve accuracy, reduce overfitting, and make more robust predictions. The idea behind RF is that by training multiple decision trees on different subsets of the data, the overall model becomes more generalized, as each tree may capture different patterns and relationships in the data (Hu and Szymczak, 2023). The following diagram shows the specific flow of the random forest algorithm.

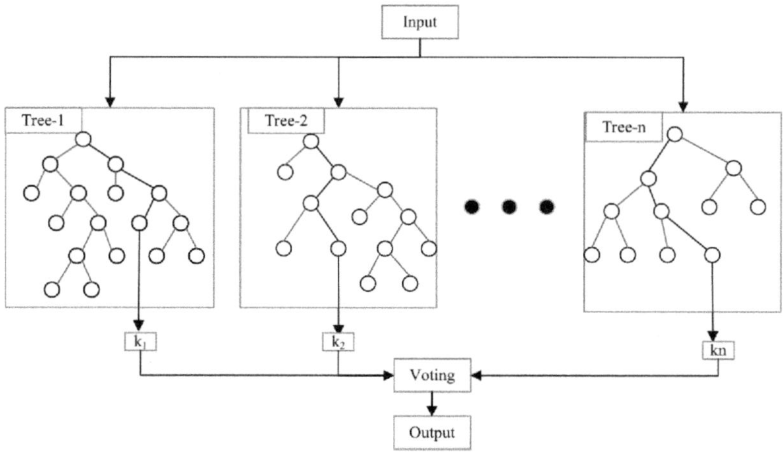

Fig. 3.2 Flow of random forest algorithm.

The RF model can be described mathematically as follows:

For Regression: Let the prediction of a single decision tree be $h_i(x)$, where $i = 1, 2, ..., N$ (the number of trees in the forest) and x is the input feature vector. The final prediction of the RF for regression is the average of all the individual tree predictions (Eq.(3.11)):

$$\hat{y} = \frac{1}{N}\sum_{i=1}^{N} h_i(x) \qquad ...(3.11)$$

Where: $\hat{y}$ is the final prediction; N is the number of trees in the forest; $h_i(x)$ is the prediction of the i-th tree.

For Classification: In classification tasks, each tree provides a prediction for a class label, and the final prediction is made by majority voting. Let $C_i(x)$ be the predicted class of the i-th tree. The final class prediction is the class with the most votes across all trees (Eq. (3.12)):

$$\hat{y} = \text{mode}\,(C_1(x), C_1(x), ..., C_N(x)) \qquad ...(3.12)$$

Where, $\hat{y}$ is the predicted class label; $C_i(x)$ is the class predicted by the i-th tree; N is the total number of trees.

b. Applications in Transportation

Traffic accidents are usually caused by the influence of people, cars, roads, and the environment. Based on the excellent performance of the RF algorithm, we prioritized the influence of traffic accident factors on the severity of traffic accidents. We used the data of automobile traffic accidents from 2018 to 2020 in the Chinese National Automobile Accident In-Depth Investigation System. The prediction features of traffic accident severity are innovated. Four accident features that did not participate in the importance ranking were added, namely, accident location, accident form, road information, and collision speed. Eight accident features (engine capacity, hour of day, age of vehicle, month of year, day of week, age band of drivers, vehicle maneuver, and speed limit) have been used in previous studies. RF was used to rank the importance of the 12 accident features, and 7 important accident features were finally adopted. By comparing the algorithms and optimizing the results, the prediction model of traffic accident degree with higher accuracy is obtained.

c. Data Description and Processing

The data are from 2800 automobile traffic accidents collected by the Chinese National Automobile Accident In-Depth Investigation System (NAIS) from 2018 to 2020. NAIS refers to the traffic accident database of NHTSA in the United States and GIDAS in Germany and combines the characteristics of traffic accidents in China. It is jointly established by the Defective Product Management Center of AQSIQ of China, together with a number of university vehicle accident research institutions and judicial appraisal institutions. The accident data include the basic variables of traffic accidents, i.e., time, location, age, gender, and degree of injury. The data collection area is distributed in the Northeast, North, East, South, Southwest, and Central regions.

The dataset is simply analyzed by statistical analysis. We found that some accident cases had incomplete accident characteristic data and some obviously wrong accident characteristic information. The method of direct deletion was used to deal with it. If the value is obviously abnormal, it is also removed directly. For example, drivers in China are only allowed to drive if they are 18 years old or older. For 0.6% of traffic accidents under the age of 18, the law was deleted. After data processing, 2,800 groups of traffic accident data were obtained.

d. Model Setup

The importance of parameters in a RF model can vary based on the specific dataset and context, but generally, the following parameters are often considered most influential in determining the model's performance:

- n_estimators: Increasing the number of trees generally improves model performance and stability, but it also increases computation time.
- max_depth: This is critical for controlling overfitting. Properly setting the maximum depth of the trees helps to balance bias and variance.

For our classification task, Accuracy, Recall, F1 score, and Roc were chosen to evaluate the model's performance comprehensively.

e. Main Result

All data were randomly selected from 2240 traffic accidents to plant forests. The remaining 560 accidents were tested. The importance of features is measured by looking at tree nodes that use this feature to reduce impurities in all trees in the forest. The final importance ranking is shown in Figure 3.3.

Through the selected 7 characteristics of traffic accident data, we conducted several experiments on the original data of 2800 traffic accidents. 2240 traffic accident data were randomly selected for training, and the rest data were used as test sets. RF, BP neural network, SVM, and radial basis neural network were used for training and testing. The prediction performance of each model on the severity of traffic accidents is shown in Table 3.5.

The performance of the RF algorithm is better than the other models. The recall is higher than that in the other models, and the false alarm rate is lower than that in the other models, so the overall F1 score is also higher. The highest ROC indicates that the prediction result of this model is more reliable and stable.

From the traffic safety management perspective, more accurate prediction of the severity of traffic accidents has always been an important research area that helps maintain a safe and efficient traffic system (Zhang et al., 2022). Many traffic accidents happen accidentally before people can take more effective measures to

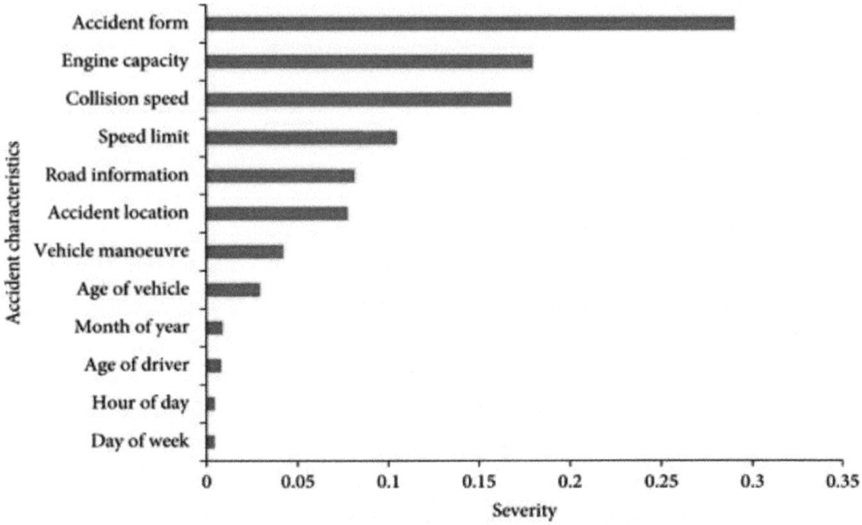

Fig. 3.3 Importance of accident characteristics.

Table 3.5 Prediction performance of each model

Model	*Accuracy (%)*	*Recall*	*F1 score*	*ROC*
Random forests	80	0.83	0.82	0.85
RBF neural network	63	0.66	0.66	0.70
BP neural network	60	0.65	0.62	0.62
Support vector machine	59	0.62	0.60	0.61

reduce the damage. The prediction model based on the RF algorithm presented in our research has good performance in terms of the severity of traffic accidents. It provides an effective reference for drivers and traffic safety management departments to proactively prevent the occurrence and minimize the impact of traffic accidents. For example, some key measures include the enhanced design of the vehicle body structure, the improvement of the road surface condition, the improvement of the road speed limit, and the setting of warning signs in high-risk road sections.

3.4.3 Gradient Boosting Regression Trees (GBRT)

a. Methodology

Gradient Boosting Regression Trees (GBRT) is a powerful ensemble learning technique, used primarily for regression tasks, though it can also be adapted for classification. The core idea behind GBRT is to combine multiple weak learners (typically decision trees) to create a strong predictive model by iteratively minimizing the prediction error. Below is an explanation of the methodology step-by-step:

(1) Ensemble Learning: GBRT is an ensemble method, meaning it aggregates the predictions of several base learners (decision trees in this case) to make a final prediction. Instead of building all trees independently (as in a RF), GBRT builds them sequentially, with each tree trying to correct the errors made by the previous trees.
(2) Weak Learners (decision trees): The weak learners in GBRT are typically shallow decision trees (also called decision stumps). These trees are weak in the sense that they perform only slightly better than random guessing, but when combined in an ensemble, they perform much better.
(3) Gradient Descent: GBRT uses a gradient descent approach to minimize the loss function (the difference between the predicted values and the actual values). Each subsequent tree is trained to predict the residual errors (gradients) of the previous model.

Gradient Descent Steps:

- Step 1: Start with an initial prediction for each training example. Usually, this is a constant value like the mean of the target variable.
- Step 2: Compute the residuals, which are the differences between the actual target values and the predicted values from the current model. Residuals represent the error the model still needs to learn to correct.
- Step 3: Train a new decision tree to predict these residuals (errors).
- Step 4: Add the predictions of this new tree to the overall model's predictions, effectively updating the model.
- Step 5: Repeat this process iteratively, with each new tree focusing on learning the residuals (gradients) from the current combined model.

Mathematically, this approach is approximated using a fitting function, which is represented as a linear combination of multiple base learners, as described in Eq. (3.13).

$$f(x) = \sum_{m=1}^{M} f_m(x) = \sum_{m=1}^{M} \lambda_m \tau(x; \sigma_m) \qquad ...(3.13)$$

Let $\tau(x; \sigma_m)$ denotes the m-th base learner with multiplier λ_m and parameter σ_m, parameter is computed by fitting the negative gradient of the loss with respect to the previous model's predictions, and the multiplier is determined by minimizing the overall loss function, as described in Eqs. (3.14) and (3.15).

$$\sigma_m = \arg\min_{\sigma} \sum_{i=1}^{N} \left(-\left[\frac{\partial L(y_i, F(x_i))}{\partial F(x_i)} \right]_{F(x_i)=f_{m-1}(x_i)} - \tau(x; \sigma_m) \right)^2 \qquad ...(3.14)$$

$$\lambda_m = \arg\min_{\lambda} \sum_{i=1}^{N} L(y_i, F(x_i) + \lambda_m \tau(x; \sigma_m)) \qquad ...(3.15)$$

Herein, $L(y_i, F(x_i))$ represents the loss function, y_i is the actual value of the dependent variable, and N is the sample size. The first term in Eq. (3.14) corresponds to the negative gradient. To prevent overfitting and underfitting, a learning rate $\eta(0 < \eta < 1)$ is introduced to enhance the model. It improves the model's generalization ability. When the model reaches the predetermined number of iterations M, the final fitted function can be expressed by Eq (3.16).

$$f_M(x) = f_{M-1}(x) + \eta\lambda_M\tau(s; \sigma_M) \qquad ...(3.16)$$

b. Applications in Transportation and Logistics

Urban freight transportation is crucial for modern societies, supporting commercial activities and supply chains. However, it also brings significant environmental challenges, especially air pollution. Diesel trucks, commonly used in urban freight transportation and logistics, are major sources of harmful emissions, e.g., nitrogen oxides (NOx) and fine particulate matter (PM2.5). These emissions constitute a large part of total vehicle emissions (Ministry of Ecology and Environment of the People's Republic of China, 2021), worsening urban air quality. Combined with the urban heat island effect, these pollutants accumulate in the city air, posing significant health risks to urban residents (Agarwal and Tandon, 2010). Some compounds in NOx also act as greenhouse gases, adding to the pressure of climate change and global warming. Despite significant efforts to promote vehicles with clean energy sources, private cars remain the primary mode of transportation in major cities, hindering the shift from diesel trucks. As urban freight demand is expected to grow, pollutant emissions are likely to increase.

To address these issues, it is essential to accurately assess the distribution of urban freight pollutant emissions, analyze key influencing factors, and propose targeted solutions from an urban planning or traffic control perspective. For this purpose, the GBRT model is employed due to its effectiveness in capturing non-linear relationships, absence of pre-assumptions about variable interactions, ability to provide insights into variable importance, and robustness to multicollinearity. GBRT has been widely used in transportation studies to explore non-linear relationships (Peng et al., 2023).

c. Data Description and Processing

We collected trajectory data from 1,460 truck drivers in Chengdu from December 16, 2018, to January 15, 2019, with a sampling frequency of 0.2Hz, totaling 28,046,480 trajectory points. Each point records the vehicle's unique identifier, latitude and longitude, speed, date, and time. Additionally, we gathered information on truck length, weight, and the type of goods through the online freight platform's shipment data, linked with the vehicle's unique identifier. Due to the differences in data sources, 4.25% of vehicles could not be matched with shipment data and were subsequently excluded.

To facilitate a more detailed examination of emissions within the city and analyze the relationship between urban infrastructure and freight-related emissions, we standardized the unit of measurement and divided the study area into a 1 km × 1 km grid. This approach offers several advantages: high spatial resolution helps generate higher gradients, which are beneficial for regression models; the 1 km grid allows for the direct conversion of quantities into densities, providing more intuitive insights; and the grid size is consistent with existing studies, facilitating comparison and validation. This division resulted in a total of 14,825 grids, enabling us to capture variations in emissions more accurately across the city. We then used the International Vehicle Emission (IVE) model, which is widely used for calculating vehicle emissions, to compute the emissions at each trajectory point and aggregate these emissions to the grids, serving as the dependent variable in this study.

The built environment and urban freight emissions have a complex interrelationship. Elements, e.g., traffic design, land use patterns, and population density in cities not only profoundly influence the distribution of urban freight demand and the structure of supply chains but also indirectly determine the efficiency and methods of freight activities, thereby significantly impacting pollutant emissions. For example, a high-density urban environment may promote short-distance transportation, reducing total travel distance and thus lowering emissions. Besides, diverse and well-designed urban environments can improve transport efficiency, reduce traffic congestion, and further decrease pollutant emissions. Based on these complex interactions, this study selected 24 relevant variables from multiple key aspects as independent variables to systematically analyze the specific impact of the built environment on freight-related pollutant emissions. These variables cover aspects such as density (e.g., building density, population density), diversity (e.g., land use mix), and design (e.g., street network, public facility layout), aiming to comprehensively reveal how characteristics of the built environment affect urban freight transportation and logistics activities and their environmental impacts. All variables involved in this study are listed in Table 3.6.

Table 3.6 Variable description

Variables	*Description*	*Unit*
Independent variable		
Traffic Design		
Main Road Density	The length of the main road within the grid	meter
Secondary Road Density	The length of the main road within the grid	meter
Main Road Intersections Count	The number of intersections between main roads in the grid.	/
Main and Secondary Intersections Count	The number of intersections between main and secondary roads in the grid.	/

Contd.

Variables	*Description*	*Unit*
Diversity and Density		
Industrial Land Ratio	The proportion of industrial land within the grid.	/
Residential Land Ratio	The proportion of residential land within the grid.	/
Commercial Land Ratio	The proportion of commercial land within the grid.	/
Public Land Ratio	The proportion of public land within the grid.	/
Transportation Facilities Land Ratio	The proportion of land within the grid dedicated to transportation facilities.	/
Land-use mix	Entropy index of POI land type in the grid.	/
Built-up Area	Urban built-up area within the grid.	km^2
NDVI Index	The normalized difference vegetation index within the grid	/
Distance		
Average Distance to logistics hubs	Average distance from the grid to various logistics hubs.	meter
Distance nearest to logistics hubs	Distance from the grid to the nearest logistics hub.	meter
Demand		
Population	Average population within the grid.	persons
GDP	Average gross domestic product within the grid.	million yuan
Housing Price	Average housing price within the grid.	Yuan/ m^2
Electricity Consumption	Average electricity consumption within the grid.	kWh
Residential POI Count	Number of Residential POIs within the grid.	
Factory POI Count	Number of Industrial POIs within the grid.	/
Restaurant POI Count	Number of restaurant POIs within the grid.	/
Logistics-related POI Count	Number of logistics-related POIs within the grid.	/
Commercial Office POI Count	Number of commercial office POIs within the grid.	/
Consumption-related POI Count	Number of consumption-related POIs within the grid.	/
Independent variable		
NOx & PM2.5 concentrations	The content values of PM2.5 and NOx caused by trucks within the grid.	g

d. Model Setup

We used the aforementioned independent and dependent variables as inputs for the model, with the model samples being 14,825 grids. To optimize the model's hyperparameters, a grid search method was employed. The final hyperparameter combination was set to a maximum of 500 regressors, a tree depth of 4, and a learning rate of 0.05. Furthermore, we used the Ordinary Least Squares model (OLS) as a baseline for comparison.

e. Main Result

After completing the training of the model, the GBRT model was able to provide the relative importance of the variables, while the OLS model allowed us to assess the impact of independent variables on the dependent variables through standardized regression coefficients, and to determine the direction of influence based on the sign of the coefficients. The results are summarized in Table 3.7.

Table 3.7 Results of the models

Variable	*GBRT*		*OLS*	
	Relative importance	*Cumulative importance*	*Estimate coefficient*	*Influence direction*
Main Road Density	41.16%	41.16%	0.33	Positive
Built-up Area	11.57%	52.73%	0.1	Positive
Industrial Land Ratio	7.50%	60.23%	0.15	Positive
Electricity Consumption	7.42%	67.65%	0.15	Positive
Housing Price	7.08%	74.73%	–0.07	Negative
Logistics-related POI Count	5.98%	80.70%	0.15	Positive
Distance to the nearest logistics hub	5.09%	85.79%	–0.1	Negative
Land-use mix	4.22%	90.01%	0.06	Positive
NDVI Index	2.86%	92.87%	–0.01	Negative
Consumption-related POI Count	2.24%	95.11%	0.04	Positive
Population	1.43%	96.54%	0.02	Positive
Restaurant POI Count	1.03%	97.57%	–0.02	Negative
Secondary Road Density	0.58%	98.15%	0.05	Positive
Public Land Ratio	0.35%	98.50%	0.01	Positive
GDP	0.29%	98.79%	0.01	Positive
Factory POI Count	0.19%	98.98%	0.04	Positive
Transportation Facilities Land Ratio	0.18%	99.15%	0.03	Positive
Residential Land Ratio	0.17%	99.32%	0.03	Positive
Average Distance to the logistics hub	0.16%	99.48%	0.03	Positive

Contd.

Variable	*GBRT*		*OLS*	
	Relative importance	*Cumulative importance*	*Estimate coefficient*	*Influence direction*
Commercial Land Ratio	0.15%	99.62%	0.04	Positive
Residential POI Count	0.12%	99.74%	0.03	Positive
Commercial Office POI Count	0.10%	99.84%	–0.06	Negative
Main Road Intersections Count	0.09%	99.93%	0.01	Positive
Main and Secondary Intersections Count	0.07%	100.00%	0.01	Positive
R^2		0.9042		0.461

In Table 3.7, variables are ranked by importance, with the top ten accounting for over 95% of the total importance, highlighting their significant impact on freight-related pollutant emissions. The GBRT model's R^2 value shows its superior suitability for capturing pollutant trends.

Most variables are statistically significant, emphasizing their impact on emissions. Main road density accounts for 41.16% of the study's importance, highlighting its critical role in freight transportation and logistics activities and pollutant emissions. Secondary roads provide density that is less important (<1%).

In the OLS model, built-up area, industrial land ratio, and electricity consumption show significant positive correlations, indicating a close relationship between freight pollution and land use, development intensity, and energy consumption. Variables such as housing price, distance to the nearest logistics hub, and NDVI index exhibit negative correlations, reflecting urban characteristics and greenness levels. Specific POIs related to logistics and consumer activities highlight their economic significance in freight-related pollutant emissions.

GBDT can also use interpretable methods such as Partial Dependence Plots (PDP) to observe the specific impact of particular independent variables on the dependent variable. Figure 3.4 shows one of the results, and more results are available in the reference (Peng et al., 2024).

Figure 3.4(a) shows the PDP relationship between main road density and freight-related pollutant emissions. As main road density increases, both NOx and PM2.5 emissions rise, peaking around 9560 m/km^2. High-density areas (>9560 m/km^2) in Chengdu's city center, with frequent truck flows, exhibit the highest PM2.5 emissions.

Figure 3.4(b) highlights that these high-density areas are concentrated in Chengdu's main urban area, along major routes. This pattern underscores the importance of strategic land use planning to minimize pollution impacts, such as by separating high-traffic areas from residential, park, and educational lands.

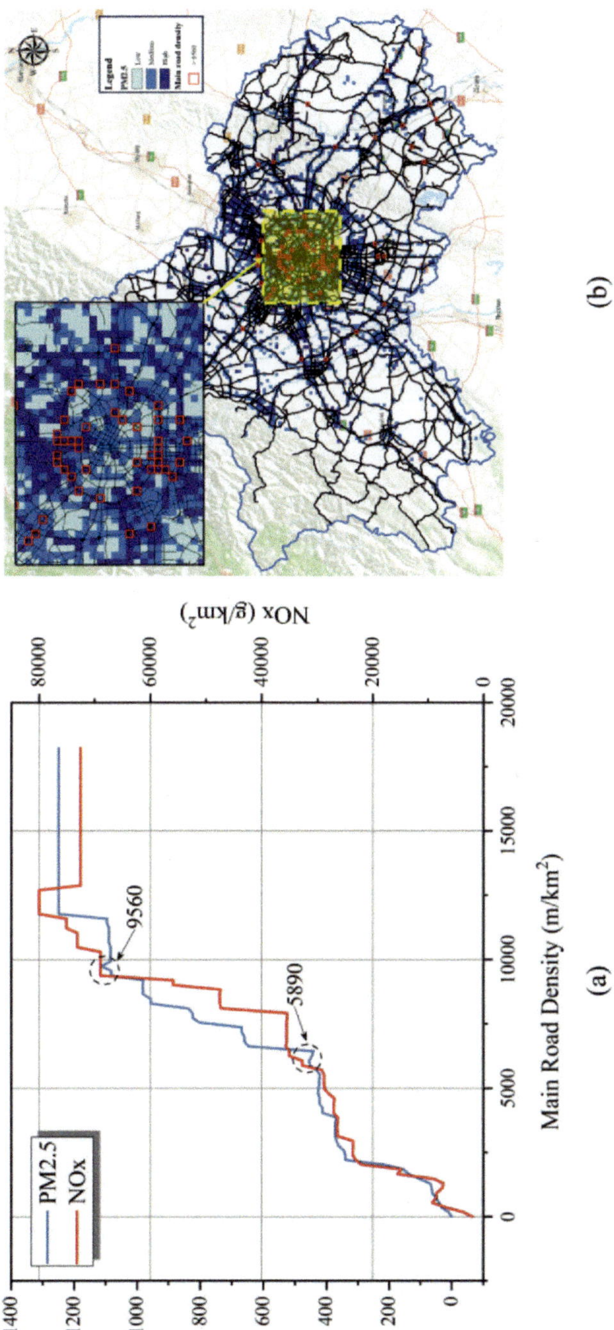

Fig. 3.4 The Relationship between main road density and freight-related pollutant emissions.

3.5 Concluding Remarks

This chapter provides a comprehensive overview of AI and data-driven smart transportation and logistics, with a focus on the theoretical foundations and applications of unsupervised learning methods and ensemble learning methods. For unsupervised learning methods, GMM is introduced in detail, and its performance is compared with several other algorithms. For ensemble learning methods, RF and GBRT are thoroughly discussed to show their applications in data-driven smart transportation and logistics.

In particular, a case study is provided to investigate the relationship between urban built environment factors and pollutant emissions from freight activities using the GBRT model. The conclusions obtained have significant implications for urban traffic management and planning. First, decision-makers gain valuable data support, aiding in the formulation of targeted policies that balance urban development with environmental protection. Second, by deepening the understanding of pollution sources and influencing factors, specific recommendations can be offered for urban planning to enhance living quality and foster cleaner and healthier urban environments.

Declaration of Generative AI and AI-assisted Technology in the Writing Process

During the preparation of the manuscript, Qwen and ChatGPT were used to polish the scientific writing. After using these tools, the authors reviewed and edited the content as needed and take full responsibility for the content of the paper.

Acknowledgment

The research is supported by the Industry 5.0 enabled Smart Logistics: A Global Perspective project funded by the Norwegian Directorate for Higher Education and Skills (HK-dir) under the Utforsk Programme (UTF-2021/10166), the National Science Foundation of China (52372306, and 52472333) and the International Cooperation Plan under the Department of Science and Technology in Sichuan Province (No.2024YFHZ0315).

References

Agarwal, M. and Tandon, A., 2010. Modeling of the urban heat island in the form of mesoscale wind and of its effect on air pollution dispersal. Applied Mathematical Modelling 34, 2520–2530. https://doi.org/10.1016/J.APM.2009.11.016.

Al-Sai, Z.A., Abdullah, R. and Husin, M.H., 2019. Big Data Impacts and Challenges: A Review, in: Jaber, K.M. (Ed.), 2019 IEEE Jordan International Joint Conference On Electrical Engineering And Information Technology (JEEIT). Presented at the IEEE Jordan International Joint Conference on Electrical Engineering and Information Technology (JEEIT), IEEE, New York, pp. 150–155.

Aziz, R., Kedia, M., Dan, S., Basu, S., Sarkar, S., Mitra, S., Mitra, P., 2016. Identifying and Characterizing Truck Stops from GPS Data, in: Perner, P. (Ed.), Advances in Data Mining. Applications and

Theoretical Aspects, Lecture Notes in Computer Science. Springer International Publishing, Cham, pp. 168–182. https://doi.org/10.1007/978-3-319-41561-1_13

Bai, X., Ma, Z. and Zhou, Y., 2023. Data-driven static and dynamic resilience assessment of the global liner shipping network. Transportation Research Part E: Logistics and Transportation Review 170, 103016. https://doi.org/10.1016/j.tre.2023.103016

Ben Ayed, A., Ben Halima, M. and Alimi, A.M., 2015. Big Data Analytics for Logistics and Transportation, in: 2015 4th IEEE International Conference on Advanced Logistics and Transport (ICALT). Presented at the 4th IEEE International Conference on Advanced Logistics and Transport (IEEE ICALT), IEEE, New York, pp. 286–291.

Chen, B.Y., Chen, X.-Y., Chen, H.-P., Huang, Y.-B., Jia, T. and Lam, W.H.K., 2024. Understanding user equilibrium states of road networks: Evidence from two Chinese mega-cities using taxi trajectory mining. Transp. Res. Pt. A-Policy Pract. 180, 103976. https://doi.org/10.1016/j.tra.2024.103976

Chen, C.Y.T., Sun, E.W., Chang, M.-F. and Lin, Y.-B., 2023. Enhancing travel time prediction with deep learning on chronological and retrospective time order information of big traffic data. Ann Oper Res. https://doi.org/10.1007/s10479-023-05223-7.

Chung, S.-H., 2021. Applications of smart technologies in logistics and transport: a review. Transportation Research Part E: Logistics and Transportation Review 153, 102455. https://doi.org/10.1016/j.tre.2021.102455.

Deng, J., Cui, Y., Chen, X., Bachmann, C. and Yuan, Q., 2023a. Who are on the road? A study on vehicle usage characteristics based on one-week vehicle trajectory data. Int. J. Digit. Earth 16, 1962–1984. https://doi.org/10.1080/17538947.2023.2218117.

Deng, J., Cui, Y., Chen, X., Bachmann, C. and Yuan, Q., 2023b. Who are on the road? A study on vehicle usage characteristics based on one-week vehicle trajectory data. International Journal of Digital Earth 16, 1962–1984. https://doi.org/10.1080/17538947.2023.2218117.

Duan, M., Qi, G., Guan, W. and Guo, R., 2020a. Comprehending and Analyzing Multiday Trip-Chaining Patterns of Freight Vehicles Using a Multiscale Method with Prolonged Trajectory Data. J. Transp. Eng., Part A: Systems 146, 04020070. https://doi.org/10.1061/JTEPBS.0000392.

Duan, M., Qi, G., Wei, G. and Guo, R., 2020b. Comprehending and Analyzing Multiday Trip-Chaining Patterns of Freight Vehicles Using a Multiscale Method with Prolonged Trajectory Data. J. Transp. Eng. Pt A-Syst. 146, 04020070. https://doi.org/10.1061/JTEPBS.0000392.

Dugalam, R. and Prakash, G., 2024. Development of a random forest based algorithm for road health monitoring. Expert Syst. Appl. 251, 123940. https://doi.org/10.1016/j.eswa.2024.123940.

Gong, L., Yamamoto, T. and Morikawa, T., 2018a. Identification of activity stop locations in GPS trajectories by DBSCAN-TE method combined with support vector machines. Transportation Research Procedia 32, 146–154. https://doi.org/10.1016/j.trpro.2018.10.028.

Gong, L., Yamamoto, T. and Morikawa, T., 2018b. Identification of activity stop locations in GPS trajectories by DBSCAN-TE method combined with support vector machines, in: Bonnel, P., Munizaga, M., Morency, C., Trepanier, M. (Eds.), Transport Survey Methods in the Era of Big Data: Facing the Challenges, Transportation Research Procedia. Presented at the 11th International Conference on Transport Survey Methods, Elsevier Science Bv, Amsterdam, pp. 146–154. https://doi.org/10.1016/j.trpro.2018.10.028.

Goswamy, A., Abdel-Aty and M., Islam, Z., 2023. Factors affecting injury severity at pedestrian crossing locations with Rectangular RAPID Flashing Beacons (RRFB) using XGBoost and random parameters discrete outcome models. Accident Analysis & Prevention 181, 106937. https://doi.org/10.1016/j.aap.2022.106937.

He Y., Gan M., Qian Q. and Liu X., 2024. Research on Intercity Truck Travel Preferences Based on Trip Chain Perspective. Frontiers of Science and Technology of Engineering Management 1–11.

Hu, J. and Szymczak, S., 2023. A review on longitudinal data analysis with random forest. Brief. Bioinform. 24. https://doi.org/10.1093/bib/bbad002.

Kumar, D., Singh, R.K. and Mishra, R., Vlachos, I., 2024. Big data analytics in supply chain decarbonisation: a systematic literature review and future research directions. International Journal of Production Research 62, 1489–1509. https://doi.org/10.1080/00207543.2023.2179346.

Li wen, 2024. Research on Logistics Transportation and Transportation Data Mining Driven by Big Data. China Shipping Gazette 55–57.

Mansoor, U., Jamal, A., Su, J., Sze, N.N. and Chen, A., 2023. Investigating the risk factors of motorcycle crash injury severity in Pakistan: Insights and policy recommendations. Transp. Policy 139, 21–38. https://doi.org/10.1016/j.tranpol.2023.05.013.

Mao J.-L., Wu T., Li S.-J., Guo Y., Zhou A.-Y., In C.-Q. and Qian W.-N., 2023. Data-driven Logistics Map Building for Bulk Commodity Transporting: Architecture and Research Progress. Journal of Software 34, 421–443. https://doi.org/10.13328/j.cnki.jos.006434.

Meng, H., Tong, X., Zheng, Y., Xie, G., Ji, W. and Hei, X., 2022. Railway accident prediction strategy based on ensemble learning. Accid. Anal. Prev. 176, 106817. https://doi.org/10.1016/j.aap.2022.106817

Ministry of Ecology and Environment of the People's Republic of China, 2021. China Mobile Source Environmental Management Annual Report.

Mohammed, A. and Kora, R., 2023. A comprehensive review on ensemble deep learning: Opportunities and challenges. J. King Saud Univ.-Comput. Inf. Sci. 35, 757–774. https://doi.org/10.1016/j.jksuci.2023.01.014.

Osaba, E., Sanchez Medina, J.J., Vlahogianni, E.I., Yang, X.-S. and Masegosa, A.D., Perez Rastelli, J., Del Ser, J., 2020. Data-driven optimization for transportation logistics and smart mobility applications [guest editorial]. IEEE Intell. Transport. Syst. Mag. 12, 6–9. https://doi.org/10.1109/MITS.2020.3017033.

Peng, T., Chen, J., Liu, K., Qiu, Z., Fu, Z. and Huang, Y., 2023. Examining the relationship between built environment and urban parking demand from the perspective of travelers. Journal of Cleaner Production 385, 135766. https://doi.org/10.1016/j.jclepro.2022.135766.

Peng, T., Gan, M., Yao, Z., Yang, X. and Liu, X., 2024. Nonlinear impacts of urban built environment on freight emissions. Transportation Research Part D: Transport and Environment 134, 104358. https://doi.org/10.1016/j.trd.2024.104358.

Shozi, N.A., Mtsweni, J., 2016. Big Data Privacy and Security: A Systematic Analysis of Current and Future Challenges, in: Proceedings of the 11th International Conference on Cyber Warfare and Security (ICCWS 2016). Presented at the 11th International Conference on Cyber Warfare and Security (ICCWS), Acad Conferences Ltd, Nr Reading, pp. 296–303.

Tsang, Y.P., Choy, K.L., Wu, C.H., Ho, G.T.S., Lam, H.Y. and Tang, V., 2018. An intelligent model for assuring food quality in managing a multi-temperature food distribution centre. Food Control 90, 81–97. https://doi.org/10.1016/j.foodcont.2018.02.030.

Wang, T., Zheng, Z., Rehmani, M.H., Yao, S. and Huo, Z., 2019. Privacy Preservation in Big Data From the Communication Perspective-A Survey. IEEE Commun. Surv. Tutor. 21, 753–778. https://doi.org/10.1109/COMST.2018.2865107.

Wang, Z., Goodchild, A.V. and McCormack, E., 2016. Freeway truck travel time prediction for freight planning using truck probe GPS data. European Journal of Transport and Infrastructure Research. https://doi.org/10.18757/EJTIR.2016.16.1.3114.

Wu, W., Xia, Y. and Jin, W., 2021. Predicting Bus Passenger Flow and Prioritizing Influential Factors Using Multi-Source Data: Scaled Stacking Gradient Boosting Decision Trees. IEEE Trans. Intell. Transp. Syst. 22, 2510–2523. https://doi.org/10.1109/TITS.2020.3035647.

Yang, Y., Jia, B., Yan, X.-Y., Chen, Y., Song, D., Zhi, D., Wang, Y. and Gao, Z., 2023. Estimating intercity heavy truck mobility flows using the deep gravity framework. Transportation Research Part E: Logistics and Transportation Review 179, 103320. https://doi.org/10.1016/j.tre.2023.103320.

Yang, Y., Shao, Z., Hu, Y., Mei, Q., Pan, J., Song, R. and Wang, P., 2022. Geographical spatial analysis and risk prediction based on machine learning for maritime traffic accidents: A case study of Fujian sea area. Ocean Eng. 266, 113106. https://doi.org/10.1016/j.oceaneng.2022.113106.

Yao, Z., Gan, M., Qian, Q., Qiao, Y. and Wei, L., 2023. Variation of truck emission by trip purposes: cases by real-world trajectory data. Transportation Research Part D: Transport and Environment 122, 103887. https://doi.org/10.1016/j.trd.2023.103887.

Yao, Z. and Yi, W., 2014. License plate detection based on multistage information fusion. Inf. Fusion 18, 78–85. https://doi.org/10.1016/j.inffus.2013.05.008.

Zafarzadeh, M., Wiktorsson, M., Baalsrud Hauge, J., 2021. A systematic review on technologies for data-driven production logistics: their role from a holistic and value creation perspective. Logistics 5, 24. https://doi.org/10.3390/logistics5020024.

Zhang, Y., Li, H. and Ren, G., 2022. Estimating heterogeneous treatment effects in road safety analysis using generalized random forests. Accid. Anal. Prev. 165, 106507. https://doi.org/10.1016/j.aap.2021.106507.

Zheng Q., Gan M., Yao Z. and Wei L., 2024. Research on Deep Learning Model for Location of High-speed Railway Freight Stations. Journal of the China Railway Society 46, 36–45.

Zhu, D., 2018. IOT and big data based cooperative logistical delivery scheduling method and cloud robot system. Futur. Gener. Comp. Syst. 86, 709–715. https://doi.org/10.1016/j.future.2018.04.081.

4 Artificial Neural Network Methods in Smart Logistics: Concept, Methods, and Applications

Mi Gan[1*], Zijun Guo[2], Zihan Zhang[3], Qian Zheng[4], Tao Peng[5] and Hao Yu[6*]

4.1 Artificial Neural Networks (ANNs)

Artificial Neural Networks (ANNs) are computational models inspired by the human brain's neural networks, designed to process and analyze data through interconnected neurons. Since the 1950s, ANNs have been widely applied in fields of pattern recognition, predictive analytics, and automatic control, significantly advancing the development of artificial intelligence (Zador, 2019).

An ANN typically consists of an input layer, hidden layers, and an output layer, each containing multiple neurons. Neurons receive input signals, process them through weighted sums and activation functions, and pass the results to the next layer. Activation functions introduce non-linear factors, enabling the network to handle complex patterns and relationships. Common activation functions include Sigmoid, ReLU, and Tanh.

Training of ANNs primarily relies on the backpropagation algorithm, which aims to minimize a loss function to enhance the model's prediction accuracy. Various optimization algorithms, i.e., Gradient Descent, Adam, and RMSProp, are employed to adjust the weights and biases during training. To prevent overfitting, several techniques, e.g., regularization, Dropout, and cross-validation, are often utilized. During the training phase, the network learns from extensive datasets, continually refining its parameters to improve performance on unseen data.

ANNs encompass various architectures tailored for specific tasks. Understanding these common types is crucial for selecting the appropriate model for a given problem. Based on the characteristics of different learning layers of ANN and their application scenarios, network structures can be divided into four types: FNN, RNN, CNN, and GAN (Abdolrasol et al., 2021).

- *Feedforward Neural Networks (FNNs):* This is the simplest form of neural network, where the information flows in one direction—from the input layer, through the hidden layers, to the output layer. FNNs are primarily used

[1-5] School of Transportation and Logistics, Southwest Jiaotong University, Chengdu, China.
[6] Department of Industrial Engineering, UiT—The Arcticl University of Norway, Narvik, Norway.
* Corresponding author: migan@swjtu.edu.cn, hao.yu@uit.no

DOI: 10.1201/9781003466451-4

for static pattern recognition tasks, e.g., image classification and function approximation. They consist of layers of neurons, and each neuron in one layer is connected to every neuron in the next layer. The absence of cycles in this architecture simplifies the training process and allows for efficient computation.

- *Recurrent Neural Networks (RNNs):* RNNs are designed for sequence data, where the output from previous steps is fed back into the network as input for the current step. This feature enables RNNs to maintain a form of memory, making them suitable for tasks such as natural language processing, time series forecasting, and speech recognition. However, traditional RNNs struggle with long sequences due to the vanishing and exploding gradient problems, which make it difficult to learn long-term dependencies in the data.
- *Convolutional Neural Networks (CNNs):* CNNs are specifically designed to process data with a grid-like topology, e.g., images. They use convolutional layers to automatically detect and learn spatial hierarchies of features. CNNs are characterized by their ability to reduce the number of parameters through weight sharing and pooling layers, which summarize the features present in the data. This architecture is highly effective for tasks involving image recognition, object detection, and video analysis.
- *Generative Adversarial Networks (GANs):* GANs consist of two neural networks, namely, the generator and the discriminator, which compete against each other (Wang et al., 2017). The generator creates synthetic data, while the discriminator evaluates its authenticity against real data. This adversarial process leads to the generator producing increasingly realistic data. GANs have gained popularity in image generation, style transfer, and data augmentation.

Table 4.1 provides an overview of the four types of ANNs.

Table 4.1 Overview of the four types of ANNs

Algorithms	*Pros and cons*	*Applications*
Feedforward Neural Networks (FNNs)	*Pros*: Easy to implement, suitable for problems where data relationships are straightforward. *Cons:* Lacks memory, unable to effectively model sequential or time-dependent data.	FNNs perform well in system optimization, such as in traffic light control and rocket trajectory optimization, by optimizing network structure and weights to enhance performance (Takayama et al., 2000; Zhao et al., 2021).
Recurrent Neural Networks (RNNs)	*Pros:* Suitable for sequential data, can model temporal dependencies, captures patterns over time. *Cons:* Issues with vanishing gradients, limited long-term memory, computationally intensive.	RNNs are commonly used for analyzing time-series data such as predicting traffic energy consumption, traffic flow, and carbon emissions estimation (Mahjoub et al., 2022; Hussain et al., 2021; Jin et al., 2022).

Contd.

Algorithms	*Pros and cons*	*Applications*
Convolutional Neural Networks (CNNs)	*Pros:* Highly effective for processing visual data, captures spatial features well, especially useful in tasks with strong local features. *Cons*: Not ideal for handling sequential or temporal data, high computational cost for training.	CNNs are widely used in visual analysis and image processing, such as in medical image diagnosis, forest fire smoke detection, and human detection in emergency search and rescue operations (Heidari et al., 2020; Zhao et al., 2021; Sambolek and Ivasic-Kos, 2021).
Generative Adversarial Networks (GANs)	*Pros:* Can generate new data from existing datasets, useful for creating synthetic scenarios, helps overcome data imbalance. *Cons:* Difficult to train, sensitive to hyperparameter tuning, potential mode collapse.	GANs have the ability to generate realistic synthetic data, making them particularly suitable for data augmentation across various fields, such as noise removal in medical imaging, and joint generation of 2D and 3D images (Zhang et al., 2020; Li and Wand, 2016).

The rest of the chapter is organized as follows. Section 4.2 provides a comprehensive literature review on the applications of ANN methods in smart logistics. Section 4.3 introduces the long short-term memory methods and their applications in smart logistics. Section 4.4 explores ANN-based computer vision technologies and their applications to enhance logistics operations. Finally, Section 4.5 summarizes the paper.

4.2 ANN Methods in Smart Logistics

ANNs are mathematical models inspired by biological neural networks, which are widely used in smart logistics due to their powerful capabilities in handling nonlinear and complex systems. In the management of logistics systems, the involvement of numerous influencing factors, complex data structures and characteristics, and high dynamism and uncertainty make traditional linear models unsuitable for addressing these challenges. ANN, on the other hand, with its adaptive learning, parallel processing, and nonlinear mapping characteristics, can effectively manage these complexities.

Demand forecasting is an essential aspect of smart logistics management. By learning from historical and real-time data, ANN can enhance the accuracy of demand forecasting, helping decision-makers better cope with uncertainty. Li et al. (2022a) developed an indicator system for regional logistics demand forecasting influenced by e-commerce and the pandemic, using an ANN model to predict the logistics demand in Changsha, China. The results indicate that the proposed ANN network outperforms the other methods including the Grey Model (1,1) and linear regression in terms of prediction accuracy, showcasing ANN's effectiveness in reducing prediction errors and enhancing the prediction of regional logistics demand. Similarly, in passenger flow forecasting, ANN has also shown precise prediction capabilities. Li et al. (2020) proposed a Spatio-Temporal Structured LSTM (ST-LSTM) model combining multi-source data for short-term prediction of origin-destination (OD) matrices in rail transit systems. In this research, smart card data, weather data, and mobile phone data were used to extract 31 features incorporating spatiotemporal characteristics, which were used as model inputs. To further enhance prediction accuracy, the hidden layer and neuron structure of the standard LSTM were improved, further enhancing the model's ability to capture the spatiotemporal dynamics of rail transit passenger flow.

In freight transportation network planning and management, ANN methods support facility location, logistics management, and freight transportation network planning. For instance, using big data, Zheng et al. (2024) developed a deep learning-based method to locate high-speed rail freight stations. The study considered site selection across three dimensions: market demand, transportation, and developmental environment, using ten indicators. A model optimized with genetic algorithms was constructed, effectively learning site selection patterns and supporting a more informed decision-making process. Casabianca et al. (2021a) proposed a method using a Bidirectional LSTM (BiLSTM) with an attention mechanism to predict vehicle destinations. This study used a deep learning approach to predict vehicle destinations based on historical travel data and tested models including Dense Neural Network (DNN) and LSTM without the attention mechanism. Experimental results show that the BiLSTM model with the attention mechanism performed best in terms of prediction accuracy and demonstrated superior robustness and stability compared to the other models tested, providing valuable implications for transportation route optimization and risk prevention.

In the field of computer vision, ANN plays a crucial role in achieving automation and intelligence in smart logistics scenarios through computer vision technologies by enabling tasks such as object detection, image classification, and scene understanding. Artificial Neural Networks (ANN) excel at recognizing patterns in visual data, making them invaluable for applications like cargo recognition and sorting technology, logistics process monitoring, and quality inspection.

Several studies focused on the applications of ANN in cargo recognition and sorting. In this regard, Qi and Sun (2024) proposed a visual recognition system based on deep learning and multimodal fusion technologies for robotic arm automatic grasping of disordered and stacked express parcels. The system integrates Convolutional Neural Networks (CNNs) and Long Short-Term Memory (LSTM) networks for image feature extraction and temporal data processing, and it uses deep fusion techniques to combine data from various sensors, e.g., visual and tactile information. This approach enables the system to accurately identify the position and orientation of parcels in complex environments, adjusting the grasping strategy based on the state of the target object, thus achieving efficient automated grasping. Experimental results indicate that the designed system demonstrates robust performance in the automatic identification of parcels in the grabbing area. Liu et al. (2022) proposed a CNN-based deep learning model for precise warehouse good recognition in order to improve the efficiency of robotic picking operations. The model focuses on recognizing objects of arbitrary shapes and updates the loss function using exponential decay and sliding averages. The system utilizes visual recognition tools for real-time identification and classification of goods, facilitating the automated picking process. Based on the key features, i.e., shape, size, and material of the goods, different intelligent grasping strategies are selected to optimize the picking of various types of goods. Mu et al. (2024) developed a robot validation platform that completes precise sorting operations via visual perception, proposing a sorting method that integrates visual image acquisition, position detection, and robot control. The Im-AlexNet model was employed to process individual or stacked parcels and to obtain their centroid coordinates for guiding the robot in sorting. This method significantly improves the efficiency, accuracy, and positioning of robot sorting tasks, especially for applications in hazardous environments.

The process monitoring in smart logistics was also emphasized. For instance, El-sayed et al. (2022) proposed a package transportation solution based on a real-time camera tracking system for parcel tracking on hexagonal omnidirectional wheeled conveyors. The study used the YOLOv5 model, which demonstrates high precision and frame rate in real-time environments and drastically improves the accuracy and efficiency of package tracking. Xie and Yao (2023) proposed a logistics warehouse object tracking method based on YOLOv5 and DeepSORT, incorporating multi-object tracking algorithms. The method can effectively differentiate between humans and goods and complete tracking within 30 milliseconds. Maalouf et al. (2024) introduced a real-time robotic system named "Follow Anything" (FAn) for detecting, tracking, and following arbitrary objects in an open vocabulary environment. This approach is based on YOLO and deep learning models, combined with several foundational models, i.e., SAM, DINO, and CLIP, to recognize and segment target objects via text, image, or click queries, while accounting for occlusion and object reappearance.

In quality inspection, the application of computer vision plays an increasingly important role in cargo packaging. Mu et al. (2024) proposed a visual perception-based method for robot item classification and sorting in warehouses. This method improves image recognition accuracy, consequently enhancing sorting efficiency and reducing the risk of packaging damage. In practical applications, this technology enhances safety for the transportation of hazardous materials and chemicals. Sa et al. (2020) proposed a packaging defect detection system based on machine vision and deep learning, which uses OpenCV for preprocessing images of damaged packaging and trains with the ResNet neural network model. This system automates the detection of packaging defects, replacing manual inspection, and has great market potential due to its suitability for packaging inspections of a variety of products, e.g., foods, medicine, and cosmetics.

In summary, advancements in data acquisition and computing power are driving the expanding application of ANN methods in smart logistics, where their self-learning, adaptive, and high-dimensional data processing capabilities render them essential for the enhancement and optimization of complex, dynamic logistics networks.

4.3 Long Short-Term Memory (LSTM) Models and Their Applications

4.3.1 Long Short-Term Memory (LSTM) Models and Methods

Recurrent Neural Networks (RNNs) are designed to process sequential data by retaining previous information through internal loops. However, standard RNNs encounter challenges when dealing with long sequences due to the vanishing or exploding gradient problem (Sherstinsky, 2020). This limitation makes it difficult for the model to capture long-term dependencies, which is crucial for tasks requiring sustained memory, such as complex language understanding and extended time-series prediction.

To address these challenges, LSTM networks were introduced by Hochreiter and Schmidhuber in 1997 (Lindemann et al., 2021). LSTM models can effectively solve the gradient vanishing problem of RNNs through the incorporation of cell states and gating mechanisms. The architecture of LSTM includes the following components:

- *Cell State:* Functions like a conveyor belt, allowing information to flow unchanged over time, which preserves long-term dependencies.
- *Forget Gate:* This gate determines which information should be discarded from the cell state, thus allowing the model to forget irrelevant data.
- *Input Gate:* It decides which new information to add to the cell state, enabling the model to incorporate relevant inputs.
- *Output Gate:* This gate controls what information is sent out from the cell state, influencing the current output of the LSTM.

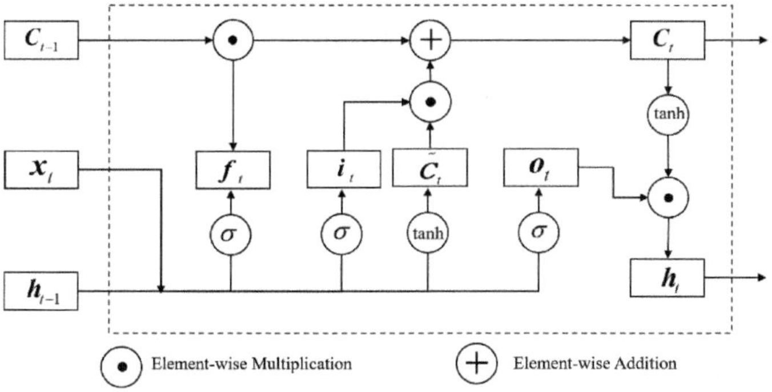

Fig. 4.1 Internal structure of the LSTM model.

Figure 4.1 shows the internal structure of an LSTM, where t is the time step index, C_t is the cell state at the current time step. In the figure, W_f, W_i, W_C, W_o, b_f, b_C, and b_o represent the weight matrices, and bias vectors for the forget gate, input gate, cell state, and output gate during network training, respectively.

The current input x_t and the hidden state from the previous time step h_{t-1} are the inputs to the current forget gate. The output, after passing through an activation function, is a vector with values ranging between 0 and 1, representing the extent to which the previous information is forgotten, as shown in Eq. (4.1).

$$f_t = \sigma(W_f [h_{t-1}, x_t] + b_f) \tag{4.1}$$

In the equations, σ represents the activation function. The input gate represents the added information, where the Sigmoid function determines the information that needs to be updated. Besides, i_t is the decay coefficient of the input gate at time step t, and a new vector $\tilde{C}_t$ is generated through the tan h function given by Eqs. (4.2) and (4.3):

$$i_t = \sigma(W_i [h_{t-1}, x_t] + b_i) \quad ...(4.2)$$

$$\tilde{C}_t = \tanh(W_C [h_{t-1}, x_t] + b_C) \quad ...(4.3)$$

The final output gate o_t and the hidden state at this time step h_t are computed as shown in Eqs. (4.4-4.6):

$$o_t = \sigma(W_o [h_{t-1}, x_t] + b_o) \quad ...(4.4)$$

$$C_t = f_t \times C_{t-1} + i_t \times \tilde{C}_t \quad ...(4.5)$$

$$h_t = o_t \times \tan h(C_t) \quad ...(4.6)$$

The advantages of LSTM include its ability to maintain long-term gradient propagation, which effectively captures long-term dependencies in data. This makes LSTMs particularly suitable for handling complex sequential information. Their wide-ranging applications in natural language processing, speech recognition, and time-series forecasting further highlight their robustness.

4.3.2 Applications in Smart Logistics

In smart logistics, LSTMs have been extensively used due to their capability to effectively manage time-series and sequential data. Applications include demand forecasting, inventory management, and transportation scheduling. For instance, Li et al. (2022b) employed LSTM models to improve regional logistics demand forecasting accuracy, while Casabianca et al. (2021b) utilized LSTM and Bidirectional LSTM networks with attention mechanisms to predict vehicle destinations based on journey history data. Nguyen et al. (2021) utilized LSTM networks to forecast multivariate time series data and detect sales anomalies in supply chain management. Additionally, Jin et al. (2022b) proposed an improved LSTM model incorporating weather influence to predict short-term parking occupancy rates, achieving better prediction accuracy of transportation-related carbon emissions.

Apart from time-series prediction tasks, LSTM networks have the ability to handle complex and nonlinear relationships in data that surpass traditional models, making them a powerful tool in other smart logistics applications, e.g., facility location problems. In this regard, LSTMs can effectively model intricate nonlinear interactions among multiple influencing factors, e.g., market demand, transportation infrastructure, and developmental environment (Zheng et al., 2024).

a. Data Collection

In this example, to address the complex heterogeneity and nonlinear relationships in the site selection of the high-speed rail freight station, multi-source data were collected to construct a set of site selection features. By utilizing several tools, i.e., ArcGIS, Python, and web crawlers, we processed and extracted features from government public data, statistical yearbooks, transportation websites, and official websites of express delivery companies. Based on the three dimensions of market demand, locational transportation, and development environment, a site selection feature set was first constructed, as shown in Table 4.2. To build the training dataset, we collected existing or planned high-speed rail freight stations from various provincial high-speed rail freight planning policies as positive samples (labeled as 1), and randomly selected ordinary high-speed rail stations at a ratio of 1:5 as negative samples (labeled as 0), resulting in a total of 576 samples.

Table 4.2 Influencing factors for the site selection of high-speed freight stations

Aspect	*Feature*	*Description*
Market Demand Conditions	Regional GDP	2021 regional GDP of the city where the site is located
	Regional Population Density	Population within a 3 km radius of the site
	Express Business Volume	Volume of express deliveries completed in 2021 in the city where the site is located
	Number of Express Service Outlets	Total number of express service outlets in the city where the site is located
Regional Traffic Conditions	Road Traffic Accessibility	Average shortest travel time from the site to other nodes in the road network
	Coverage of Station Nodes	Average daily number of trains serviced by the site
	Average Daily Service Frequency of Stations	Number of nodes where the trains serviced by the site stop
	Degree of Airport Hub Connection	Number of airports within a 200 km radius of the site

Contd.

Aspect	*Feature*	*Description*
Development Environment Conditions	City Administrative Level	Administrative level of the city where the site is located
	Electric Multiple Unit Depot Reservation	Whether the site has reserved an electric multiple unit (EMU) depot

b. Model Framework

The study adopted a deep learning model using LSTM networks as the primary learning layer, stacked with dense layers (Dense), to capture the complex nonlinear relationships between site selection features and outcomes, as shown in Figure 4.2. The input data was the standardized site selection feature matrix. To enhance model performance, we first used a genetic algorithm to optimize the model's hyperparameters, including the number of layers in the LSTM and Dense layers, the number of neurons, and the Dropout rate. Subsequently, we conducted a feature importance analysis to assess the impact of each feature on the model's predictions, which would inform feature selection and model improvement in subsequent steps. The model's output was a suitability prediction for each station, providing a label (1 indicating suitability for constructing a freight station and 0 indicating unsuitability). The model's evaluation metric was the F1 score, which comprehensively considers precision and recall.

c. Results

We calculated the importance of each feature using the Permutation Feature Importance method and ranked them from high to low. Based on this ranking, we determined 10 feature combinations; for example, the 6-feature combination includes the top six features. The model's performance was simulated under each feature combination. When the number of features is 5, the model in this study achieved the best performance, with a predicted F1 score of 0.9444, which is an improvement of 3.23 compared to that before parameter optimization (F1 score of 0.9121) and an increase of 2.3 compared to that before feature optimization (F1 score of 0.9214), resulting in an overall improvement of 5.21.

Through the Permutation Feature Importance method, we identified the main features affecting site selection, including node coverage rate, availability of EMU depots, express business volume, highway transportation accessibility, and airport hub connectivity. This helps reduce the input dimensionality of the model and improves its generalization ability. Using Zhejiang Province as an example, the model predicts 13 potential locations for constructing high-speed rail freight stations, including Hangzhou East, Hangzhou South, and Yuyao North. The results align with the government planning, which verifies the model's effectiveness.

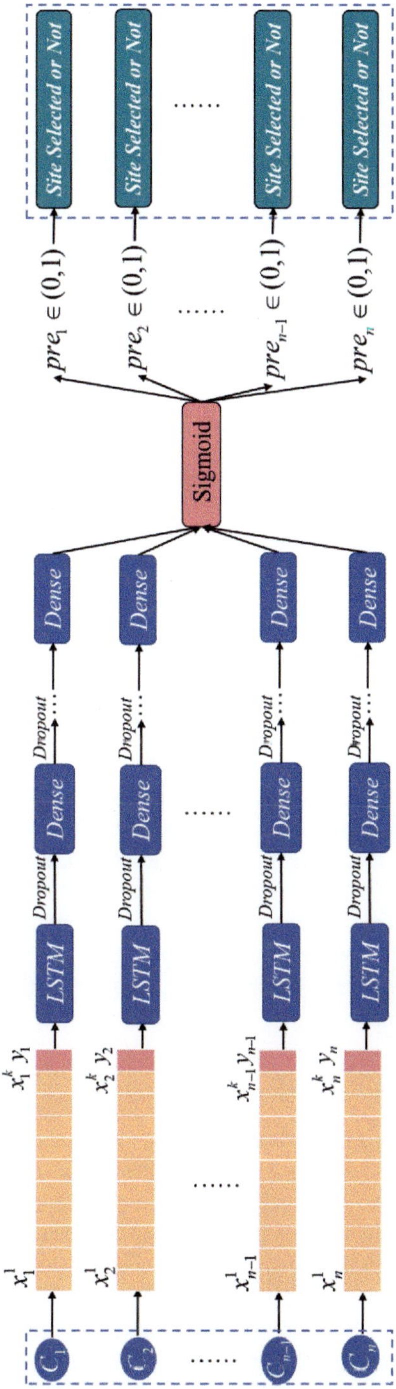

Fig. 4.2 The model structure based on LSTM for site selection.

d. Conclusion

This example illustrates how a deep learning-based method can be used for site selection of high-speed rail freight stations, which effectively captures the complex nonlinear relationships between site selection features and outcomes. Through the collection and processing of multi-source data, a site selection feature system has been established. By utilizing an LSTM model combined with a genetic algorithm, the prediction accuracy of the model can be significantly improved. The case study demonstrates that the model can provide effective decision support for the site selection of high-speed rail freight stations, consequently optimizing the regional logistics network planning.

4.4 ANN-based Computer Vision and Their Applications in Smart Logistics

In the previous section, we explored the application of ANN in data prediction, and ANN also plays a crucial role in computer vision. Since the early 1980s, neural networks have been applied to image recognition tasks. In 1986, David Rumelhart and Geoffrey Hinton introduced the backpropagation algorithm, which provided an effective method for training multilayer neural networks, thereby drastically advancing the application of ANN in visual tasks (Rumelhart et al., 1986). In 1989, Yann LeCun's research on the use of CNN for handwritten digit recognition marked an important milestone for ANN in the application of computer vision and demonstrated its exceptional capability in processing image data, particularly in the automatic extraction of visual features (LeCun et al., 1989).

The advantages of ANN in computer vision lie in its ability to automatically extract features and learn from vast datasets, significantly improving the accuracy of tasks such as image classification, object detection, and image segmentation. Compared to traditional algorithms, ANN possesses the ability to adapt and improve, thereby exhibiting higher effectiveness in dynamic environments. This adaptability provides ANN with a notable advantage when processing logistics-related images, as the logistics environment frequently undergoes changes and requires rapid adjustments to respond to new conditions.

4.4.1 Theoretical Framework

Computer vision is a field of technology that enables computers to "see" and "understand" images or videos. By simulating the human visual system, computer vision models can recognize and interpret objects, scenes, and activities within images, thereby facilitating various practical applications. In this section, we will explore the working principles of computer vision models, understand how they process input images, and outline the steps and significance of model training. Thus, this section provides a theoretical foundation for subsequent applications.

The functioning of models in the field of computer vision primarily involves the following key steps:

- *Image Input and Preprocessing:* Computer vision systems first receive images or video frames, typically represented as two-dimensional pixel matrices. To enhance the model's efficiency and accuracy, the input images undergo a series of preprocessing steps, including normalization (scaling pixel values to a specific range), resizing (adjusting the image to a fixed size required by the model), and data augmentation (such as rotation and flipping to generate additional training samples).
- *Feature Extraction and Decision Mapping:* In the core phase of the model, deep learning architectures, e.g., convolutional neural networks (CNNs), extract relevant features from images. Through convolutional layers, the model automatically learns local patterns, e.g., edges and textures. Activation functions, e.g., ReLU, introduce non-linearity, enabling the model to capture complex relationships. The extracted features are then downsampled through pooling layers to reduce the dimensionality of the data while preserving important information. Subsequently, fully connected layers map the features to the output space for specific tasks, such as class predictions in image classification or bounding box coordinates in object detection.
- *Output and Postprocessing:* The final output format of the model depends on the specific task. In image classification, it provides a probability distribution over the categories to which the image belongs; in object detection, it outputs the object categories along with their bounding box coordinates; in semantic segmentation, it predicts a category label for each pixel. To enhance the quality of the predictions, post processing is often necessary, such as applying non-maximum suppression (NMS) to eliminate overlapping prediction boxes and retain the most likely object locations.

After understanding the working principles of computer vision models, it is essential to grasp the necessity of model training. Training is the process that enables the model to learn and extract useful information from input images. Through training, the model continuously adjusts its parameters based on existing labeled data, thereby enhancing its ability to recognize and understand new images. In the following, the training steps are detailed for computer vision models.

The training steps for computer vision models typically encompass the following key phases:

- *Data Preparation and Partitioning:* Collect labeled image data to ensure data quality, and partition the dataset into training, validation, and test sets for training, hyperparameter tuning, and model evaluation.
- *Model Selection and Configuration:* Select an appropriate model architecture based on the characteristics of the data, define a loss function to measure the discrepancy between the model's output and the true labels, and choose an optimization algorithm, e.g., Adam or SGD, to update the model weights, which influences the convergence speed and performance of the model.
- *Training and Evaluation:* Update the weights using the backpropagation algorithm, monitor the model's performance on the validation set to adjust hyperparameters, and upon completion of training, evaluate the model's

generalization capability using the test set, calculating various performance metrics.

Next, the YOLOv8 model is used as an example to understand how a computer vision model based on ANN works. The YOLO (You Only Look Once) series of models represents a real-time object detection algorithm based on deep learning. Since its initial introduction in 2015, it has undergone multiple iterations and updates, with each version improving and optimizing the previous one to accommodate various application scenarios and enhance detection performance. The design philosophy of YOLOv8 aims to achieve real-time and high-precision object detection (Varghese and Sambath, 2024). It efficiently performs feature extraction and object recognition under limited computational resources, thereby meeting the application needs across various fields, e.g., automation, security monitoring, and medical image analysis.

4.4.2 The Architecture of the YOLOv8 model

a. Overall architecture

The YOLOv8 model, like other visual deep learning models, is divided into three main components: Backbone, Neck, and Head, as shown in Figure 4.3. The Backbone is responsible for feature extraction, processing the input images through a convolutional neural network to capture important patterns. Next, the Neck facilitates the fusion of multi-scale features, enhancing YOLOv8's ability to detect objects of varying sizes. At this stage, modules similar to Feature Pyramid Networks (FPN) are typically employed to strengthen multi-scale representations. Finally, the Head is responsible for generating the final prediction results, including bounding boxes and object category probabilities, ensuring that YOLOv8 can accurately and efficiently perform object detection tasks.

YOLOv8 continues the multi-scale feature detection mechanism of the YOLO series, outputting three feature maps of different sizes. This allows YOLOv8 to achieve greater flexibility and accuracy in multi-scale object detection, enabling it to effectively handle objects ranging from large to small simultaneously.

b. Loss function

The primary purpose of the loss function is to measure the discrepancy between the model's predictions and the actual data, guiding the optimization process of the model. The loss function of YOLOv8 comprises three main components:

b.1 Binary Cross-Entropy Loss (BCE Loss): This loss is used to assess the model's confidence in the presence of an object within each predicted bounding box. BCE Loss measures the difference between the predicted confidence and the true labels, encouraging the model to accurately determine the presence or absence of an object. BCE Loss is calculated by Eq. (4.7).

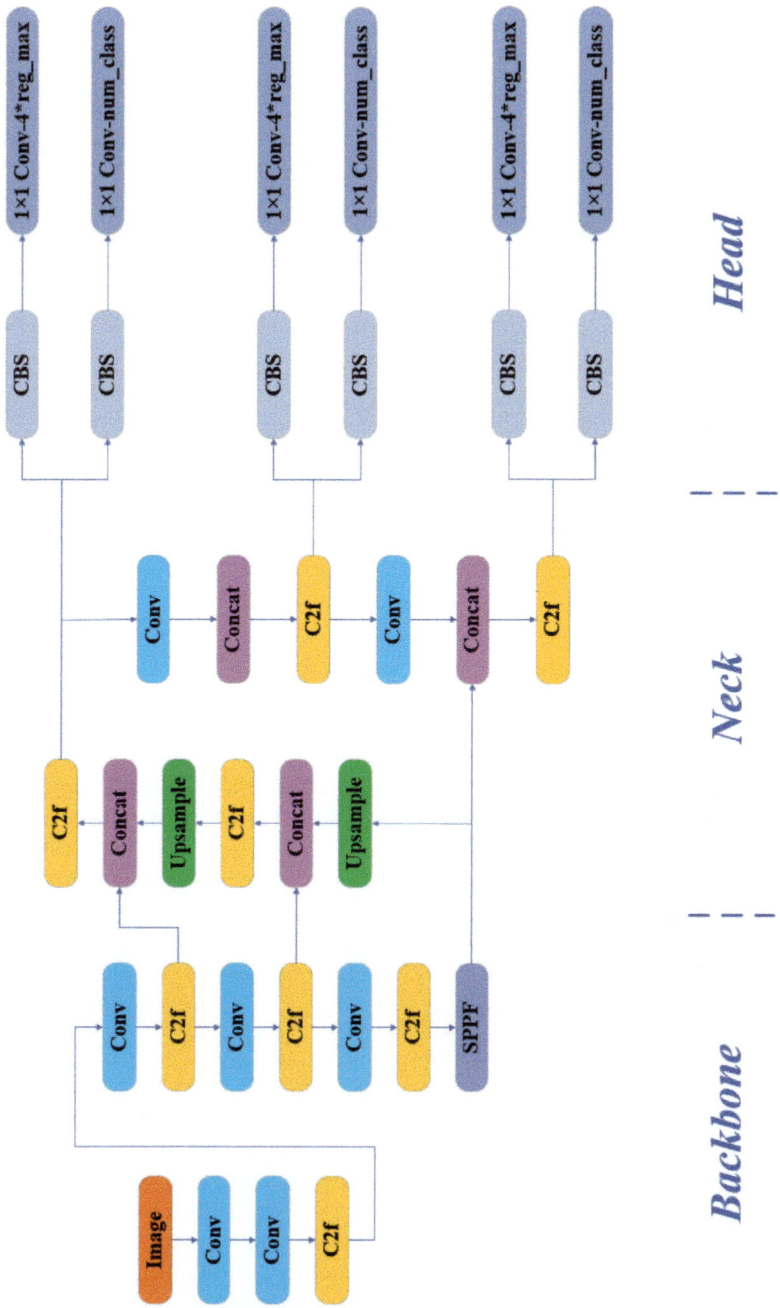

Fig. 4.3 Overall architecture of the YOLOv8 model.

$$L_{\mathrm{BCE}} = \sum_{i=1}^{N} (y_i \log(\hat{y}_i) + (1 - y_i) \log(1 - \hat{y}_i)) \qquad \text{...(4.7)}$$

Herein, y_i is the true confidence of the i-th box (where 1 indicates the presence of an object and 0 indicates its absence), and $\hat{y}_i$ is the predicted confidence of the i-th box.

b.2 Distribution Focal Loss (DFL): This loss enhances the model's focus on hard-to-classify samples by incorporating additional distributional information. The formula is provided in Eq. (4.8).

$$L_{\mathrm{DFL}} = \sum_{i=1}^{N} \sum_{j=1}^{K} p_{ij} \log(\hat{p}_{ij}) \qquad \text{...(4.8)}$$

Herein, K is the number of classes for each sample, p_{ij} is the true probability distribution of the j-th class for the i-th sample, and $\hat{p}_{ij}$ is the model's predicted probability for that class.

b.3 Complete Intersection over Union Loss (CIoU Loss): This loss evaluates the overlap between the predicted bounding box and the ground truth box, taking into account the area of overlap, the distance between their center points, and the aspect ratio, resulting in more accurate bounding box predictions. CIoU Loss is calculated by Eq. (4.9).

$$L_{\mathrm{CIoU}} = 1 - IoU + \frac{(c-u)^2}{c^2} + \alpha v \qquad \text{...(4.9)}$$

Herein, IoU represents the overlap ratio, c is the diagonal length of the minimum enclosing box, u is the distance between the center points of the predicted and ground truth boxes, v represents the difference in aspect ratios, and α is a weight parameter.

The total loss of YOLOv8 is the sum of these three components and is calculated by Eq. (4.10).

$$L_{\mathrm{total}} = L_{\mathrm{BCE}} + L_{\mathrm{DFL}} + L_{\mathrm{CIoU}} \qquad \text{...(4.10)}$$

c. Optimizer

In machine learning and deep learning, an optimizer is an algorithm used to update model parameters, with the primary goal of minimizing the loss function to enhance the model's prediction accuracy. The optimizer calculates the gradients of the loss function with respect to the model parameters and uses this gradient information to adjust the parameters, thereby gradually reducing the loss.

In YOLOv8, the optimizer employed is the Adaptive Moment Estimation (Adam) optimizer. Adam combines the advantages of momentum optimization and RMSProp, allowing for the adaptive adjustment of the learning rate for each parameter, which enhances training efficiency and model performance.

The core of the Adam optimizer lies in the estimation of the first moment (mean) and the second moment (variance) of the gradients. The updated formulas are given by Eqs. (4.11) and (4.12).

First Moment Estimation (Mean):

$$m_t = \beta_1 m_{t-1} + (1 - \beta_1)\, g_t \tag{4.11}$$

Second Moment Estimation (Variance):

$$v_t = \beta_2 v_{t-1} + (1 - \beta_2)\, g_t^2 \quad \text{...(4.12)}$$

Herein, m_t is the first moment estimate at time step t (the mean of the gradients), v_t is the second moment estimate at time step t (the mean of the squared gradients), g_t is the gradient at the current time step t, β_1 and β_2 are hyperparameters typically set to 0.9 and 0.999, respectively.

After computing the first and second moments, the parameter update formula is provided in Eq. (4.13).

$$\theta_{t+1} = \theta_t - \frac{\alpha}{\sqrt{v_t} + \varepsilon} m_t \quad \text{...(4.13)}$$

In this formula, θ_t is the parameter at time step t, α is the learning rate, typically set to a small value such as 0.001, and ε is a small constant used to prevent division by zero, typically set to 10^{-8}.

By employing the Adam optimizer, YOLOv8 can dynamically adjust the learning rate during training, accelerating convergence and improving the model's robustness. This enables YOLOv8 to adapt more rapidly to the characteristics of the data when handling complex object detection tasks, thereby enhancing detection performance.

4.4.3 Applications in Smart Logistics

ANN-based computer vision technology has numerous applications in smart logistics, e.g., monitoring, automated sorting, and cargo tracking, to name a few. However, due to the inherent limitations of the models, leveraging computer vision technology to address real-world logistics challenges often requires modifications based on specific application scenarios. Common improvement methods include data augmentation, transfer learning, model pruning and quantization, and designing more targeted feature extractors by incorporating domain knowledge. By taking YOLOv8 as an example, the applications of ANN-based computer vision technology in logistics operations are discussed.

a. Description

Small object detection has long been a significant challenge in the field of computer vision, as small targets occupy fewer pixels in images, contain limited information, and are more susceptible to noise interference. Traditional object detection algorithms often perform poorly with small targets, leading to frequent false negatives and false positives. Qin et al. (2024) designed a lightweight object detection algorithm based on YOLOv8, named DBR-YOLOv8, and applied it to warehouse management, aiming to address issues related to high computational complexity, stringent real-time requirements, cargo stacking, and missed detections of small targets in storage environments.

The DBR-YOLOv8 model mainly improves upon the YOLOv8 network structure with several key enhancements. First, it employs a Bi-level Routing Attention (BRA) mechanism to enhance the model's attention allocation to critical features, thereby improving small target detection capabilities. Second, it replaces the traditional PANet with a Bidirectional Feature Pyramid Network (BiFPN) in order to effectively integrate feature information across different scales, enhancing the model's multi-scale processing ability. Last, it introduces a Dilation-wise Residual (DWR) module in the C2f structure, which increases small target detection accuracy while effectively reducing the model's computational complexity.

- *Bi-level Routing Attention (BRA):* The BRA mechanism enhances the model's focus on input data through dynamic sparse attention. Compared to the original model, BRA utilizes a two-level attention structure that first filters out irrelevant information at a coarse level and then promotes deep interaction between key tasks at a fine-grained level. This flexible attention allocation allows the model to achieve greater precision in small object detection, significantly improving its ability to concentrate on important features.
- *Bidirectional Feature Pyramid Network (BiFPN):* The BiFPN outperforms traditional pyramid networks in feature fusion. Unlike the unidirectional FPN in the original model, BiFPN enhances the fusion of low-semantic and high-semantic features through bidirectional cross connections. This improvement increases the model's ability to detect small targets and enables better utilization of information across different scales, particularly in complex scenarios.
- *C2f-DWR Module:* The C2f-DWR module addresses the issue of channel information redundancy in the original model by combining the residual structure with dilated convolutions. Compared to traditional structures, C2f-DWR expands the receptive field while maintaining lightweight computation, allowing the model to capture multi-scale features more effectively. This design is crucial for accurately detecting objects of various sizes in complex environments, significantly enhancing the model's performance.

b. Results and Discussions

Qin et al. (2024b) investigated storage environment datasets under two scenarios: retail goods and express logistics. In the retail goods scenario, they utilized the large-scale multi-camera dataset MessyTable, created in collaboration with SenseTime and Nanyang Technological University, which includes the top ten categories with a total of 11,388 images, such as Ice Dew, Nongfu Spring, Pepsi, Coca-Cola1, Coca-Cola2, Jiaduobao, Heqizheng Iced Tea, Green Tea, and AD

Calcium Milk. In the express logistics scenario, a self-built dataset comprising 1,000 images categorized into two groups: boxes and QR codes, was used.

The performance evaluation of YOLO series networks primarily relies on Precision, Recall, and Mean Average Precision (mAP), where mAP@0.5 indicates the average precision when the IoU threshold is set at 0.5. As the mAP value increases from 0.05 to 0.95, the overall accuracy of the model also improves. The evaluation metrics involve the number of true positives (TP), false negatives (FN), and false positives (FP). Furthermore, in balancing lightweight design and performance, the model's parameter count and computational complexity (FLOPs) are crucial considerations.

Through a comparative analysis of DBR-YOLOv8 against mainstream YOLO models, DBR-YOLOv8 demonstrates superior accuracy. Specifically, its mAP@0.5 exceeds that of YOLOv5n, YOLOv7-tiny, and YOLOv8n by 3.1%, 0.8%, and 1.4%, respectively, while mAP@0.5:0.95 shows increases of 8.2%, 5.2%, and 1.4% compared to these models. In terms of computational efficiency, DBR-YOLOv8 exhibits a significant reduction in FLOPs, being 6.1G lower than YOLOv7-tiny and 1.1G lower than YOLOv8n. Additionally, the model parameters of DBR-YOLOv8 are reduced by 3.8M and 0.8M compared to YOLOv7-tiny and YOLOv8n, respectively, while its model weights are 7.8MB and 1.7MB lighter than those of YOLOv7-tiny and YOLOv8n.

4.5 Conclusions

In this chapter, the concepts and methods of ANN are introduced, and the applications of ANN methods in smart logistics planning, operations, and management are discussed. In particular, ANN methods outperform traditional algorithms in demand forecasting due to their superior ability to model complex nonlinear relationships, handle large datasets with intricate patterns, and learn adaptively from real-time data, significantly enhancing prediction accuracy and responsiveness. As one of the most powerful ANN methods, a detailed introduction of LSTM networks is given, and its applications in demand forecasting and selection of facility locations are discussed. Furthermore, ANN-based computer vision technologies and their applications for objective detection, image classification, and real-time tracking in smart logistics are also introduced. Future research on ANNs in smart logistics is expected to focus on enhancing decision-making through advanced computational technology, integrating with other Industry 5.0 technologies, e.g., Internet-of-Things (IoT), real-time information from smart sensors and devices, blockchain, etc., to fundamentally transform logistics operations and management toward digitalization, smartness, human-centricity, and sustainability.

Declaration of Generative AI and AI-assisted Technology in the Writing Process

During the preparation of the manuscript, Qwen and ChatGPT were used to polish the scientific writing. After using these tools, the authors reviewed and edited the content as needed and take full responsibility for the content of the paper.

Acknowledgment

The research is supported by the Industry 5.0 enabled Smart Logistics: A Global Perspective project funded by the Norwegian Directorate for Higher Education and Skills (HK-dir) under the Utforsk Programme (UTF-2021/10166), the National Science Foundation of China (52372306, and 52472333) and the International Cooperation Plan under the Department of Science and Technology in Sichuan Province (No.2024YFHZ0315).

References

Abdolrasol, M.G.M., Hussain, S.M.S., Ustun, T.S., Sarker, M.R., Hannan, M.A., Mohamed, R., Ali, J.A., Mekhilef, S. and Milad, A., 2021. Artificial Neural Networks Based Optimization Techniques: A Review. Electronics 10, 2689. https://doi.org/10.3390/electronics10212689.

Casabianca, P., Zhang, Y., Martínez-García, M. and Wan, J., 2021a. Vehicle destination prediction using bidirectional LSTM with attention mechanism. Sens. 21, 8443. https://doi.org/10.3390/s21248443.

Casabianca, P., Zhang, Y., Martínez-García, M. and Wan, J., 2021b. Vehicle Destination Prediction Using Bidirectional LSTM with Attention Mechanism. Sensors 21, 8443. https://doi.org/10.3390/s21248443.

El-sayed, M.E., Youssef, A.W., Shehata, O.M., Shihata, L.A. and Azab, E., 2022. Computer vision for package tracking on omnidirectional wheeled conveyor: Case study. Engineering Applications of Artificial Intelligence 116, 105438. https://doi.org/10.1016/j.engappai.2022.105438.

Heidari, M., Mirniaharikandehei, S., Khuzani, A.Z., Danala, G., Qiu, Y. and Zheng, B., 2020. Improving the performance of CNN to predict the likelihood of COVID-19 using chest X-ray images with preprocessing algorithms. Int. J. Med. Inf. 144, 104284. https://doi.org/10.1016/j.ijmedinf.2020.104284.

Hussain, B., Afzal, M.K., Ahmad, S. and Mostafa, A.M., 2021. Intelligent traffic flow prediction using optimized GRU model. IEEE Access 9, 100736–100746. https://doi.org/10.1109/ACCESS.2021.3097141.

Jin, B., Zhao, Y. and Ni, J., 2022a. Sustainable transport in a smart city: prediction of short-term parking space through improvement of LSTM algorithm. Appl. Sci. 12, 11046. https://doi.org/10.3390/app122111046.

Jin, B., Zhao, Y. and Ni, J., 2022b. Sustainable Transport in a Smart City: Prediction of Short-Term Parking Space through Improvement of LSTM Algorithm. Applied Sciences 12, 11046. https://doi.org/10.3390/app122111046.

LeCun, Y., Boser, B., Denker, J.S., Henderson, D., Howard, R.E., Hubbard, W. and Jackel, L.D., 1989. Backpropagation applied to handwritten zip code recognition. Neural computation 1, 541–551.

Li, C. and Wand, M., 2016. Precomputed real-time texture synthesis with markovian generative adversarial networks, in: Computer Vision – ECCV 2016. Presented at the European Conference on Computer Vision, Springer, Cham, pp. 702–716. https://doi.org/10.1007/978-3-319-46487-9_43

Li, D., Cao, J., Li, R. and Wu, L., 2020. A spatio-temporal structured LSTM model for short-term prediction of origin-destination matrix in rail transit with multisource data. IEEE Access 8, 84000–84019. https://doi.org/10.1109/ACCESS.2020.2991982.

Li, Y. and Wei, Z., 2022a. Regional logistics demand prediction: a long short-term memory network method. Sustainability 14, 13478. https://doi.org/10.3390/su142013478.

Li, Y. and Wei, Z., 2022b. Regional Logistics Demand Prediction: A Long Short-Term Memory Network Method. Sustainability 14, 13478. https://doi.org/10.3390/su142013478.

Lindemann, B., Maschler, B., Sahlab, N. and Weyrich, M., 2021. A survey on anomaly detection for technical systems using LSTM networks. Computers in Industry 131, 103498. https://doi.org/10.1016/j.compind.2021.103498.

Liu, H., Zhou, L., Zhao, J., Wang, F., Yang, J., Liang, K. and Li, Z., 2022. Deep-learning-based accurate identification of warehouse goods for robot picking operations. Sustainability 14, 7781.

Maalouf, A., Jadhav, N., Jatavallabhula, K.M., Chahine, M., Vogt, D.M., Wood, R.J., Torralba, A. and Rus, D., 2024. Follow Anything: Open-set detection, tracking, and following in real-time. IEEE Robotics and Automation Letters 9, 3283–3290.

Mahjoub, S., Chrifi-Alaoui, L., Marhic, B. and Delahoche, L., 2022. Predicting energy consumption using LSTM, multi-layer GRU and drop-GRU neural networks. Sens. 22, 4062. https://doi.org/10.3390/s22114062.

Mu, L., Quan, C., Ren, J., Zhou, Y. and Zhao, T., 2024. Enhancing small parcel sorting accuracy: Robot machine vision in stacking target image experiment. Journal of Computational Methods in Science and Engineering 14727978241299513.

Nguyen, H.D., Tran, K.P., Thomassey, S. and Hamad, M., 2021. Forecasting and Anomaly Detection approaches using LSTM and LSTM Autoencoder techniques with the applications in supply chain management. International Journal of Information Management 57, 102282. https://doi.org/10.1016/j.ijinfomgt.2020.102282.

Qi, W. and Sun, X., 2024. The Study on Recognition and Detection of Express Package Grabbing Based on Machine Vision. Academic Journal of Science and Technology 9, 198–203.

Qin, X., Yuan, Q., Yang, E., Yang, D. and Huo, J., 2024a. YOLOv8 lightweight target detection algorithm for warehousing environment, in: 2024 IEEE 6th Advanced Information Management, Communicates, Electronic and Automation Control Conference (IMCEC). Presented at the 2024 IEEE 6th Advanced Information Management, Communicates, Electronic and Automation Control Conference (IMCEC), IEEE, Chongqing, China, pp. 877–881. https://doi.org/10.1109/IMCEC59810.2024.10574958.

Qin, X., Yuan, Q., Yang, E., Yang, D. and Huo, J., 2024b. YOLOv8 lightweight target detection algorithm for warehousing environment, in: 2024 IEEE 6th Advanced Information Management, Communicates, Electronic and Automation Control Conference (IMCEC). Presented at the 2024 IEEE 6th Advanced Information Management, Communicates, Electronic and Automation Control Conference (IMCEC), IEEE, Chongqing, China, pp. 877–881. https://doi.org/10.1109/IMCEC59810.2024.10574958.

Rumelhart, D.E., Hinton, G.E. and Williams, R.J., 1986. Learning representations by back-propagating errors. nature 323, 533–536.

Sa, J., Li, Z., Yang, Q. and Chen, X., 2020. Packaging defect detection system based on machine vision and deep learning, in: 2020 5th International Conference on Computer and Communication Systems (ICCCS). IEEE, pp. 404–408.

Sambolek, S. and Ivasic-Kos, M., 2021. Automatic person detection in search and rescue operations using deep CNN detectors. IEEE Access 9, 37905–37922. https://doi.org/10.1109/ACCESS.2021.3063681

Sherstinsky, A., 2020. Fundamentals of Recurrent Neural Network (RNN) and Long Short-Term Memory (LSTM) network. Physica D: Nonlinear Phenomena 404, 132306. https://doi.org/10.1016/j.physd.2019.132306.

Takayama, K., Morva, A., Fujikawa, M., Hattori, Y., Obata, Y. and Nagai, T., 2000. Formula optimization of theophylline controlled-release tablet based on artificial neural networks. J. Controlled Release 68, 175–186. https://doi.org/10.1016/S0168-3659(00)00248-0.

Varghese, R. and Sambath, M., 2024. YOLOv8: A Novel Object Detection Algorithm with Enhanced Performance and Robustness, in: 2024 International Conference on Advances in Data Engineering and Intelligent Computing Systems (ADICS). IEEE, pp. 1–6.

Wang, K., Gou, C., Duan, Y., Lin, Y., Zheng, X. and Wang, F.-Y., 2017. Generative adversarial networks: introduction and outlook. IEEE/CAA J. Autom. Sinica 4, 588–598. https://doi.org/10.1109/JAS.2017.7510583.

Xie, T. and Yao, X., 2023. Smart logistics warehouse moving-object tracking based on yolov5 and deepsort. Applied Sciences 13, 9895.

Zador, A.M., 2019. A critique of pure learning and what artificial neural networks can learn from animal brains. Nat Commun 10, 3770. https://doi.org/10.1038/s41467-019-11786-6.

Zhang, X., Jian, W., Chen, Y. and Yang, S., 2020. Deform-GAN:an unsupervised learning model for deformable registration. https://doi.org/10.48550/arXiv.2002.11430.

Zhao, E., Liu, Y., Zhang, J. and Tian, Y., 2021. Forest fire smoke recognition based on anchor box adaptive generation method. Electronics 10, 566. https://doi.org/10.3390/electronics10050566.

Zheng Q., Gan M., Yao Z. and Wei L., 2024. Research on Deep Learning Model for Location of High-speed Railway Freight Stations. Journal of the China Railway Society 46, 36–45.

5 Embedding Social Value in Sustainable Transport and Smart Logistics Systems

Al-Amin Abba Dabo[1*], F. Frank Alparslan[2], Emmanuel Vanger[3] and Bright Amajuoyi[4]

5.1 Introduction

Sustainability in logistics and transport has traditionally focused on environmental issues such as reducing emissions, minimising waste, and resource optimisation. While essential, this focus often neglects the equally important social dimension of sustainability. Social value—which includes community well-being, inclusion, and equity—is crucial for creating truly sustainable logistics systems. Environmental goals alone are insufficient; a broader approach that integrates social and economic aspects is required to ensure transport systems promote social equity and access to essential services like employment, education, and healthcare (Banister & Hickman, 2013). Embedding social value into logistics systems is therefore vital for establishing not only efficient but also equitable and inclusive networks (Prokopenko et al., 2024).

With the advent of Industry 5.0, the concept of sustainability in logistics is evolving. Unlike Industry 4.0, which primarily emphasised automation and efficiency, Industry 5.0 focuses on human-centred systems, encouraging collaboration between technology and people to achieve broader societal benefits. This shift presents an opportunity to address social issues such as public safety, community well-being, and economic inclusion. As Gumzej (2021) observes, Industry 5.0 envisions systems designed not just for markets but also for wider communities, providing equitable access to goods and services for marginalized populations, thus addressing long-standing disparities.

As such, technological advancements in smart logistics including the Internet of Things (IoT), Artificial Intelligence (AI), and Blockchain, offer promising ways to embed social value into transport systems. These innovations improve efficiency and address social and environmental goals by enabling real-time tracking, optimising routes, and managing resources. For instance, AI-powered route optimisation ensures essential goods reach underserved communities efficiently and sustainably, while IoT allows for dynamic adjustments to meet

[1-4] Department of Business, Systems and Operations, University of Northampton, Northampton, United Kingdom.
* Corresponding author: Al-Amin.Dabo@northampton.ac.uk

DOI: 10.1201/9781003466451-5

the needs of both urban and rural areas (Chung, 2021; Nathanail et al., 2016). Similarly, blockchain enhances transparency and accountability, ensuring ethical sourcing and distribution of goods. This is particularly crucial in sectors like pharmaceuticals and food, where public trust relies on verified and ethical supply chains (Bajak, 2022). By leveraging these technologies, logistics systems can generate both economic value and positive social impact, supporting broader objectives of inclusion and sustainability.

Consequently, this chapter explores how logistics systems can integrate social value to benefit the environment, economy, and communities. Through theoretical frameworks, key concepts, and a case study, it highlights the transformative potential of embedding social value into sustainable transport and logistics systems. As Industry 5.0 continues to shape the future, integrating social value into logistics is not just an ethical choice but a strategic necessity for building resilient and inclusive societies (Adel and Alani, 2024).

5.2 Understanding Social Value in the Context of Sustainable Transport and Smart Logistics

Sustainable transport and smart logistics are increasingly recognised as essential components of both economic and social development. While the environmental benefits of sustainable practices in these sectors, such as emissions reduction and resource efficiency, are well documented, the concept of social value remains underexplored. Social value refers to the broad range of benefits that societies gain from operations, going beyond the purely financial or economic impacts to include the social and community outcomes generated by those activities (Prokopenko et al., 2024). In the context of transport and logistics, social value can manifest in multiple ways, including improving public access to transportation, creating equitable job opportunities, reducing congestion, enhancing community safety, and ensuring that marginalized groups have access to affordable, reliable transport systems.

5.2.1 Defining Social Value

Social value refers to the broader societal benefits created by an activity, extending beyond direct economic gains to encompass positive impacts on well-being, equity, and social inclusion. In transport and logistics, social value is realised by ensuring that systems are accessible, inclusive, and affordable for all segments of society, especially those who are disadvantaged or marginalised. For instance, socially valuable transport networks provide affordable and reliable services that enable access to essential resources such as education, employment, and healthcare. This approach addresses social inequalities by prioritising the needs of individuals with disabilities, elderly citizens, and low-income populations (Nathanail et al., 2016).

In logistics, social value fosters economic development through job creation, particularly in disadvantaged areas with limited employment opportunities.

By prioritising local hiring, logistics companies stimulate economic growth, providing sustainable livelihoods for underserved communities. Furthermore, optimising transport routes and employing smart logistics technologies not only reduces emissions but also alleviates traffic congestion, enhancing public safety and improving urban quality of life (Garg et al., 2024). Ultimately, integrating social value in sustainable transport systems contributes to community well-being, promoting not just economic development but also safer, more inclusive, and equitable environments (Adel and Alani, 2024).

5.2.2 *The Link Between Social Value and Sustainable Development Goals (SDGs)*

The United Nations' Sustainable Development Goals (SDGs) offer a global framework that unites social, environmental, and economic sustainability, providing a comprehensive approach to development that the transport and logistics sectors can significantly support. Key SDGs align directly with the principles of social value in logistics, highlighting the role of transport systems in promoting equitable, inclusive, and sustainable communities (Adel and Alani, 2024). For instance, SDG 3 (Good Health and Well-being) relates closely to transport initiatives that focus on reducing traffic-related injuries, improving air quality, and providing safe, accessible means of transportation. By prioritising public health through enhanced transport infrastructure, logistics systems contribute substantially to community well-being, making essential services more accessible and fostering healthier urban environments (Prokopenko et al., 2024; Sharma et al., 2022).

SDG 8 (Decent Work and Economic Growth) and SDG 10 (Reduced Inequality) underscore the need to create fair employment opportunities, promote safe working conditions, and ensure economic stability. These goals also emphasise reducing disparities in access to essential resources, including transportation. Transport and logistics systems are pivotal in achieving these objectives by generating jobs across supply chains, infrastructure projects, and transportation services, thereby fostering economic growth and stability. Furthermore, socially inclusive logistics systems bridge inequalities by providing affordable and reliable transportation options for marginalised and low-income communities, advancing social inclusion and equal access to opportunities (Nathanail et al., 2016).

Complementing these goals, SDG 11 (Sustainable Cities and Communities) prioritises the development of transport systems that are inclusive, safe, resilient, and sustainable. By adopting smart logistics and sustainable transport strategies, cities can reduce traffic congestion, improve mobility for all residents, and create urban environments that cater to the needs of vulnerable populations. Aligning logistics operations with these SDGs enables transport providers and authorities to expand social value, fostering resilient, inclusive communities while addressing broader sustainability challenges (Shee et al., 2021; Blasi et al., 2022).

Overall, embedding social value into transport and logistics aligns well with the SDG framework, enabling logistics systems to promote equitable access to essential services, stimulate job creation, and support sustainable urban development. Through these contributions, socially responsible logistics systems become vital assets in building inclusive, resilient communities, affirming that the integration of social value into logistics is a powerful mechanism for advancing global sustainability goals.

5.3 Embedding Social Value in Sustainable Transport Systems

5.3.1 The Role of Transport in Enhancing Social Value

Transport systems play a crucial role in supporting economic development, social interaction, and access to essential services, both in urban and rural areas. Sustainable transport systems that integrate social value principles are designed to meet the needs of diverse populations, particularly underserved or marginalised communities, by prioritising accessibility, affordability, safety, and equity. By addressing these aspects, transport systems contribute to societal goals such as reducing inequality and promoting social inclusion (Oviedo & Guzmán, 2020). Public transport is especially significant in advancing social value, as it provides affordable and reliable mobility options that benefit those with limited resources. For individuals in low-income areas, public transport facilitates access to employment, education, and healthcare, thereby improving life prospects and contributing to a more equitable distribution of opportunities (Jauregui-Fung, 2021).

In addition to public transport, infrastructure that promotes active travel—such as pedestrian paths and bicycle lanes—contributes to both public health and environmental sustainability. Encouraging walking and cycling not only reduces traffic congestion and pollution but also promotes physical health by encouraging active lifestyles (Biehl et al., 2019). Non-motorised forms of transport reduce reliance on cars, leading to lower emissions and quieter, safer streets, particularly in densely populated areas. By supporting active travel, transport systems help create urban environments that are healthier, more sustainable, and socially inclusive (Lemoine et al., 2016). These benefits extend to the wider community, as such infrastructure not only reduces environmental impact but also fosters social interaction and cohesion.

The integration of technology further enhances the social value of transport systems, making them more accessible and responsive to the needs of a broad user base. Real-time information services, digital payment systems, and mobile applications offer significant improvements in the convenience and usability of public transport (Visan et al., 2022). For example, real-time data on schedules and delays allows passengers to plan journeys more efficiently, while digital payment options reduce barriers for those without access to traditional banking services (García-Suarez et al., 2018). By incorporating these technologies, transport systems can adapt to the diverse needs of their users, ensuring that they are accessible, user-friendly, and reliable.

In combining these elements, sustainable transport systems play a multi-dimensional role in enhancing social value. By providing safe, accessible, and environmentally friendly transport options, they support economic participation and contribute to the health and well-being of communities. This comprehensive approach to transport infrastructure recognises that mobility is fundamental not only for individual progress but also for achieving a more inclusive, resilient society where transport is accessible to all, regardless of socioeconomic status.

5.3.2 Social Value in Freight Transport and Urban Logistics

The concept of social value is also highly relevant to freight transport and urban logistics, particularly in densely populated urban areas. Freight systems that prioritise social value aim to minimise the negative impacts of logistics activities on urban communities. Issues such as traffic congestion, noise pollution, and environmental degradation can negatively affect the quality of life for city residents. Sustainable logistics systems seek to reduce these impacts by employing smart logistics technologies that optimise delivery routes, reduce emissions, and promote efficient resource use (Cruz Daraviña, 2018).

Technologies like real-time routing, delivery optimisation, and the use of electric or hybrid vehicles help minimise the negative externalities of freight transport, contributing to the overall well-being of urban communities (Jauregui-Fung, 2021). For example, by optimising delivery schedules to avoid peak traffic times, logistics companies can reduce congestion and improve the flow of goods without disrupting the daily lives of residents. Moreover, the use of quieter, electric delivery vehicles reduces noise pollution, creating a more peaceful and liveable urban environment. By embedding social value into freight transport and logistics, cities can become more sustainable and socially inclusive, aligning with broader goals of urban sustainability (Gumzej, 2023).

5.3.3 Case Study: Socially Inclusive Public Transport in Bogotá, Colombia

The TransMilenio Bus Rapid Transit (BRT) system in Bogotá, Colombia, stands as a globally recognised model of sustainable urban transport, designed to address the city's critical transportation challenges and provide efficient, safe, and inclusive public transport to millions of residents. The system was conceptualised in the late 1990s in response to Bogotá's rapid population growth and the simultaneous decline of its public transportation system. Bogotá, home to over 8 million residents, experienced significant internal migration during the 1990s, which exacerbated the inefficiencies of an already overburdened and underfunded public transport network (Index Mundi, 2017). Prior to the implementation of TransMilenio, Bogotá's public transport consisted of 15,000 buses operated by 66 private companies. The system was chaotic, with no designated bus stops, low service quality, and a lack of coordination, resulting in high accident rates, long commuting times, and severe air pollution (Perdomo Calvo & Mendoza, 2007).

Recognising the need for an overhaul, Bogotá's mayor, Enrique Peñalosa, in 1999, proposed a comprehensive plan for a Bus Rapid Transit (BRT) system—TransMilenio. The project aimed to improve accessibility, efficiency, safety, and environmental sustainability in the city's public transport sector. The goals were ambitious: restrict private automobile use, expand bicycle paths, improve public spaces, and reduce air pollution (Center for Public Impact, 2017). To achieve these objectives, TransMilenio was developed as a public-private partnership, with responsibilities divided between public agencies and private bus operators. The public sector oversaw infrastructure development, planning, and system regulation, while the private sector managed bus operations and maintenance (Bocarejo & Tafur, 2013; PPP IRC, 2017).

The implementation of TransMilenio began in phases. Phase 1, which included 42 km of busways, was completed in 2000. Subsequent phases expanded the system, with an additional 42 km completed in 2006, 22 km in 2012, and 7 km in 2015. Plans for future phases include the integration of a metro train system, expanding the network by 388 km by 2024 (Santana Palacios, 2021). The system is composed of large-capacity buses operating on dedicated bus lanes with elevated bus stations to facilitate faster boarding and disembarking. The use of dedicated busways and the feeder system connecting residential areas to main trunk routes ensures that the entire city, including low-income areas, has access to reliable and efficient transport (Jauregui-Fung, 2021; Pocacito, 2017).

The impact of TransMilenio on Bogotá's urban landscape has been transformative. The system has significantly improved the quality of life for the city's residents, especially those from marginalised communities. By offering affordable, efficient, and safe transportation, TransMilenio has helped reduce travel times and enhanced access to jobs, education, and healthcare. For instance, in areas where TransMilenio operates, there has been a 92% reduction in road-related deaths, a 75% reduction in injuries, and a 79% decrease in collisions. Moreover, the system's design prioritises accessibility, making it easier for individuals with disabilities and elderly people to use the service (Lemoine et al., 2016). Additionally, the environmental benefits of TransMilenio are significant. Between 2013 and 2019, the system helped reduce annual CO_2 emissions by an estimated 578,918 tons, equivalent to removing over 123,000 cars from the roads each year. This reduction in emissions has been coupled with a decrease in noise pollution, further improving urban living conditions (García-Suarez et al., 2018). The reduction in air pollutants, such as particulate matter (PM) and nitrogen oxides (NOx), has contributed to better respiratory health for Bogotá's residents, many of whom previously suffered from pollution-related illnesses (Santana Palacios, 2021).

The success of TransMilenio, however, has not come without challenges. As the system has expanded, issues such as overcrowded buses, service delays, and high fares have emerged. During rush hours, it can take up to 20 minutes to board a bus at some stations due to overcrowding, and the cost of fares has led to dissatisfaction among some users (Colombia Reports, 2017). Moreover, the reliance on diesel-powered buses has continued to generate pollution, as many of the older buses have not been replaced with cleaner, eco-friendly alternatives (Bocarejo & Tafur, 2013; PRI, 2015).

To address these issues, Bogotá has implemented several initiatives to improve the system's efficiency and sustainability. The city has introduced hybrid and electric buses into the fleet, with plans to further expand the use of low-emission vehicles in the coming years (PPP IRC, 2017). Efforts are also underway to integrate the BRT system with Bogotá's planned metro and cable car networks, which will further reduce congestion and improve overall service quality. Additionally, Bogotá has introduced car-free days, peak-hour vehicle restrictions, and expanded bicycle lanes to encourage the use of non-motorised transport and reduce the city's reliance on private cars (Cruz Daraviña, 2018).

The TransMilenio system has served as a model for other cities in Colombia and beyond. The success of Bogotá's BRT system has led to the development of similar projects in other Colombian cities, including Pereira and Cali, as well as in cities across Latin America (Good Practices in City Energy Efficiency, 2017). Bogotá's experience with TransMilenio has demonstrated that BRT systems can be scaled and adapted to meet the needs of cities with different population sizes, proving that such systems can provide a viable solution for improving urban mobility in developing countries (Santana Palacios, 2021). Looking forward, the challenges of improving service quality, integrating new transport modes, and transitioning to more sustainable energy sources remain. However, the continued investment in public transport and the prioritisation of social and environmental benefits over short-term gains highlight the city's commitment to ensuring that TransMilenio remains a critical component of Bogotá's sustainable urban future. The system's ongoing evolution serves as an example of how cities can successfully embed social value in public transport systems, creating safer, more inclusive, and sustainable urban environments.

5.4 Smart Logistics Systems: A Tool for Enhancing Social Value

5.4.1 Defining Smart Logistics

Smart logistics refers to the application of cutting-edge technologies like the Internet of Things (IoT), artificial intelligence (AI), and big data analytics to enhance logistics operations. These technologies collectively improve resource utilisation, reduce emissions, and accelerate delivery times, making logistics systems more efficient and sustainable. However, smart logistics is not solely focused on operational efficiency. It also contributes to creating inclusive and adaptive logistics networks that can respond effectively to diverse needs and challenges, ensuring that these systems remain resilient and capable of addressing societal demands (Gunasekaran et al., 2020).

5.4.1.1 The Role of Technology in Creating Social Value

Technological advancements in logistics go beyond boosting efficiency or cutting costs; they are increasingly pivotal in delivering social value by addressing broader societal issues such as sustainability, inclusivity, safety, and community welfare. These advancements have a tangible impact on society, enhancing the overall quality of life for individuals and communities. Advanced technologies are shaping the logistics landscape, helping to create a more connected, responsible, and efficient supply chain (McKinnon, 2018). Smart logistics systems leverage these advanced technologies to optimise efficiency, sustainability, and resilience. Below is an exploration of how key technologies are being deployed in logistics operations.

a. Internet of Things (IoT)
The Internet of Things (IoT) has transformed logistics operations by providing visibility and control across the supply chain. By leveraging IoT-powered sensors and devices, companies can achieve real-time tracking and monitoring of goods, vehicles, and infrastructure. This level of connectivity enables oversight of logistics operations, fostering smoother workflows and more informed decision-making (Lee et al., 2018; Ben-Daya et al., 2019). For instance, in warehouses, IoT-based inventory systems monitor stock levels, track product movements, and evaluate storage conditions such as temperature and humidity. These capabilities prevent issues like overstocking or shortages while minimising waste and improving responsiveness to demand fluctuations. The ability to adjust supply chain strategies dynamically based on real-time data ensures businesses can adapt swiftly to changing market conditions, thereby enhancing operational efficiency and resilience (Huang et al., 2019).

Predictive maintenance represents another critical application of IoT in logistics, addressing one of the industry's most significant challenges: equipment downtime. IoT devices continuously monitor the health of machinery and vehicles, collecting data on performance metrics such as vibration, temperature, and wear. By analysing this data, companies can identify potential failures before they occur, scheduling maintenance proactively to prevent costly delays (Kamble et al., 2019). This approach reduces unexpected breakdowns and enhances the reliability of logistics networks. Furthermore, IoT contributes to a superior customer experience by enabling real-time delivery tracking and status updates. Customers can monitor shipments in transit, receive accurate delivery estimates, and benefit from enhanced transparency, which fosters trust and satisfaction (Wang et al., 2020). This combination of operational efficiency and customer-focused improvements underscores the value IoT brings to modern logistics.

In addition to operational enhancements, IoT plays a vital role in driving sustainability and ethical practices in the logistics sector. Fleet management, for example, is significantly optimised through IoT systems that monitor fuel consumption, driver behaviour, and vehicle conditions in real time. These insights allow companies to reduce inefficiencies, lower fuel usage, and decrease greenhouse gas emissions, contributing to cleaner air and healthier communities (Gubbi et al., 2013). Moreover, IoT's integration with blockchain technology adds another layer of functionality by ensuring that the data collected from IoT devices is secure, tamper-proof, and transparent. This is particularly valuable in industries like pharmaceuticals and food, where product authenticity and ethical sourcing are paramount. Blockchain-powered IoT systems can track goods from their point of origin to their final destination, ensuring accountability, traceability, and compliance with ethical standards (Treiblmaier, 2019). This integration fosters trust among stakeholders, bolsters consumer confidence, and supports fair trade practices.

IoT also optimises logistics networks by streamlining routes, reducing idle times, and minimising detours. This not only lowers operational costs but also contributes to significant environmental benefits, such as reduced carbon emissions and fuel consumption (Hofmann & Rüsch, 2017). These efficiencies are particularly critical in urban logistics, where traffic congestion and pollution are pressing concerns. By leveraging IoT data, logistics companies can design routes that avoid high-traffic areas, saving time and reducing the environmental impact of their operations. Furthermore, IoT-enabled logistics systems are crucial in fostering smarter, more adaptable supply chains capable of responding to disruptions, such as natural disasters or market fluctuations. These systems provide the flexibility needed to maintain continuity in supply chains, ensuring that essential goods reach their intended destinations even under challenging circumstances.

b. Artificial Intelligence (AI)

Artificial Intelligence (AI) is enabling the optimisation of a wide array of processes across the supply chain. One prominent application is route optimisation, where AI analyses data such as traffic conditions, weather patterns, and delivery schedules to identify the most efficient routes for vehicles. This not only reduces fuel consumption and operational costs but also accelerates delivery times, resulting in enhanced logistics efficiency (Abirami et al., 2024). The real-time adaptability of AI in adjusting routes during unforeseen disruptions, such as road closures or sudden traffic congestion, further underscores its value. These capabilities support both cost savings and sustainability goals by minimising emissions and energy use.

AI also has a significant impact on demand forecasting. By processing historical data, market trends, and customer behaviour patterns, AI enables businesses to accurately predict demand fluctuations. This ensures the availability of essential goods while avoiding overstocking or shortages, which can lead to financial losses and waste (Choi, 2023). For example, during periods of sudden market shifts, such as those seen during the COVID-19 pandemic, AI-supported logistics systems allowed companies to maintain the flow of critical supplies like food and medical equipment. This predictive capability not only enhances business efficiency but also proves critical in times of crisis, ensuring supply chain stability and responsiveness.

In warehouse operations, AI-powered automation has redefined how repetitive tasks are handled. Tasks such as sorting, picking, and packing are performed with greater speed and accuracy by AI-driven systems, reducing labour costs while significantly boosting productivity (Sodiya et al., 2024). AI also plays a vital role in resource management, ensuring that workforce allocation aligns with operational demands. By analysing activity levels and forecasting peak periods, AI helps logistics companies optimise staffing, reduce inefficiencies, and ensure seamless operations even during high-demand scenarios (Toorajipour et al., 2021). These advancements not only improve operational efficiency but also allow human workers to focus on more strategic and fulfilling responsibilities, contributing to job satisfaction and retention.

AI's benefits are particularly evident during crises or emergencies when consistency and speed are critical. For instance, during the COVID-19 pandemic, AI-enabled logistics systems ensured the uninterrupted delivery of essential goods to vulnerable communities, highlighting AI's potential as a lifeline in challenging times (Marr, 2020). The ability to anticipate disruptions and prioritise shipments based on urgency allows AI-driven systems to maintain supply chain continuity, ensuring that even the most underserved areas receive the resources they need. Such applications demonstrate the broader societal value of AI in logistics, extending its impact beyond operational efficiency to include equity and access.

Moreover, AI plays a key role in advancing sustainability in logistics. By reducing inefficiencies in routing and warehouse management, AI systems lower fuel consumption and energy use, which translates into a smaller environmental footprint (Chen et al., 2024). Fleet performance can also be optimised using AI, which identifies areas for improvement, such as minimising idle times and maximising load capacity. These practices align logistics operations with sustainability goals, contributing to reduced greenhouse gas emissions and compliance with environmental regulations.

c. Advanced Analytics and Big Data

Within logistics, big data analytics serves as a critical enabler for operational efficiency and strategic decision-making. By processing vast amounts of information—such as real-time traffic conditions, weather forecasts, and customer preferences—big data helps organisations improve route planning and reduce delivery times (Hopkins & Hawking, 2018; Wang et al., 2020). This capability is particularly valuable in managing complex supply chains, where even minor inefficiencies can have significant ripple effects. Furthermore, big data provides real-time insights into inventory levels and demand fluctuations, allowing companies to prevent overstocking or shortages and maintain cost-efficient operations (Ivanov et al., 2019).

Big data analytics is also having a significant impact on urban planning by enabling smarter, safer, and more community-focused systems. By analysing data on traffic patterns, delivery schedules, and consumer behaviour, logistics companies and local governments can identify inefficiencies and implement solutions that enhance mobility and reduce congestion. For instance, authorities can use data-driven insights to optimize traffic flow, alleviate bottlenecks, and reduce vehicle emissions, creating greener, less congested cities (Kawasaki et al., 2023; Tao et al., 2023). These improvements benefit not only logistics operations but also the broader community by contributing to cleaner air and a more efficient urban environment.

Big data also plays a pivotal role in enhancing public safety. By analysing accident data, local authorities can pinpoint high-risk areas and implement preventive measures, such as better street lighting, improved pedestrian crossings, and clearer road signage (Li et al., 2018). These proactive strategies make cities safer for all users—whether they are walking, cycling, or driving. Logistics companies, too, benefit from big data's predictive capabilities. By forecasting demand and optimising delivery routes, they can reduce unnecessary trips, minimise fuel consumption, and enhance customer satisfaction through faster and more reliable deliveries (Wang et al., 2020). These optimisations not only lower operational costs but also contribute to a more sustainable logistics ecosystem.

Moreover, big data supports inclusive urban development by fostering greater community participation in planning processes. When insights derived from data analysis are shared with local residents, urban development projects can be tailored to reflect the community's specific needs and preferences. This approach ensures that cities grow in a harmonious and sustainable manner, benefiting both businesses and residents alike (Batty, 2018). For example, planners can identify underserved areas in terms of public transportation or delivery infrastructure and work collaboratively with stakeholders to address these gaps, creating equitable urban environments.

Overall, big data analytics is driving innovation in logistics and urban planning by offering actionable insights that improve operational efficiency, sustainability, and inclusivity. From optimising traffic systems and ensuring safer cities to enhancing inventory management and customer experiences, big data empowers stakeholders to make informed decisions that benefit both businesses and communities. Its potential to transform logistics and urban environments emphasises its role as a foundation of modern, data-driven strategies for sustainable development.

d. Blockchain Technology

Blockchain technology has become an integral component of modern logistics, offering secure and transparent data management throughout the supply chain.

By creating a decentralised and tamper-proof ledger, blockchain ensures that all transactions are accurately recorded and accessible to stakeholders. This transparency minimises the risk of fraud, errors, and disputes, fostering trust among logistics partners (Wang et al., 2020; Issaoui et al., 2020). One of the most significant innovations enabled by blockchain is the use of smart contracts—self-executing agreements with predefined terms coded into the blockchain. These contracts automate transactions and processes, reducing administrative costs, paperwork, and the time required for execution (Agrawal et al., 2023).

The traceability offered by blockchain is particularly critical in industries where product safety and authenticity are paramount. In sectors such as food and pharmaceuticals, blockchain enables companies to track every stage of a product's journey from origin to delivery, ensuring accountability and reducing risks associated with contamination or counterfeit goods (Queiroz et al., 2019). For example, pharmaceutical companies can leverage blockchain to maintain a verified chain of custody for medications, ensuring that only genuine products reach consumers. This capability is invaluable in addressing widespread issues such as counterfeit drugs, which pose significant risks to public health (Mackey & Nayyar, 2017).

Blockchain's emphasis on transparency extends beyond product safety to ethical labour practices and corporate responsibility. Companies can use blockchain to verify that suppliers comply with ethical standards, such as paying fair wages and avoiding exploitative practices like child labour. This level of accountability helps businesses align with socially responsible goals while building trust with consumers who increasingly demand ethical and sustainable practices (Kamilaris et al., 2019). By ensuring that suppliers and stakeholders meet these standards, blockchain fosters a culture of responsible corporate behaviour and fair-trade practices.

In addition to enhancing trust and accountability, blockchain supports sustainability initiatives by providing insights into inefficiencies within the supply chain. By analysing data on transportation routes, resource utilisation, and energy consumption, companies can identify areas for improvement, reduce waste, and lower their carbon footprint. For example, blockchain can reveal redundant processes or excessive transportation stages, enabling companies to adopt greener logistics practices and contribute to global sustainability goals (Saberi et al., 2019). Moreover, the transparency provided by blockchain fosters collaboration among supply chain participants, ensuring that all stakeholders adhere to shared sustainability and ethical standards. This collective accountability creates a logistics ecosystem that is not only efficient but also socially and environmentally responsible (Francisco & Swanson, 2018).

e. Cloud Computing

Cloud computing offers flexible and adaptable platforms for managing data, streamlining processes, and enabling real-time communication. One of its biggest advantages is scalability; as logistics operations grow or fluctuate, cloud systems can adjust without the need for costly infrastructure changes, ensuring continuous smooth operations (Hofmann & Rüsch, 2017). Cloud-based platforms also foster collaboration, enabling all supply chain participants, including suppliers, logistics providers, and customers, to access up-to-date information and make quicker, more informed decisions (Matenga and Mpofu, 2022).

Cloud computing enhances visibility and tracking capabilities, allowing companies to monitor shipments and inventory in real time, which improves operational efficiency and keeps customers updated with accurate delivery information (Lee et al., 2018). Additionally, cloud-based analytical tools enable logistics companies to optimize delivery routes, forecast demand, and manage inventory more effectively, resulting in cost savings and improved service quality (Wu et al., 2016).

f. Smart Transportation and Autonomous Delivery

Advancements in smart transportation technologies, including autonomous vehicles and drones, are boosting efficiency, reducing operational costs, and enhancing accessibility. These innovations are particularly significant in addressing the needs of remote or underserved areas, ensuring equitable access to goods and resources.

- *Autonomous Vehicles:* Autonomous vehicles operate without the need for human drivers. These vehicles can function continuously, providing uninterrupted service and significantly enhancing delivery productivity. This is particularly beneficial for long-distance and high-frequency delivery routes, where consistent operation is essential (Nordhoff et al., 2018). Autonomous vehicles are ideal for navigating challenging terrains, ensuring that rural and hard-to-reach communities can access essential goods (Klumpp, 2018). Moreover, they enhance safety by minimising the risk of human error, reducing accidents, and enabling companies to lower labour costs. The data collected by autonomous vehicles during their journeys further aids in route optimisation, improving delivery times and operational efficiency.
- *Drones:* Drones are becoming increasingly critical in addressing logistical challenges, particularly in remote or inaccessible areas. By bypassing conventional transportation routes, drones can deliver goods quickly and efficiently to regions where traditional delivery methods may struggle. This capability is especially critical during emergencies, such as natural disasters, where time-sensitive items like medical supplies need to reach isolated communities urgently (Otto et al., 2018). Drone usage in urban environments for last-mile delivery is increasingly popular due to their ability to navigate congestion and deliver packages swiftly. By reducing reliance on traditional delivery vehicles, drones contribute to sustainable logistics by producing fewer emissions and reducing energy consumption (Goodchild & Toy, 2018). The real-time data collected by drones through sensors further supports the optimisation of delivery routes, enhancing efficiency and cost-effectiveness (Mnyakin, 2023).

g. Robotics and Automation

Robotics and automation enhance efficiency, accuracy, and workforce productivity. In warehouses and distribution centers, robots are increasingly utilized to perform repetitive tasks such as sorting, picking, packing, and loading goods. These duties, while essential, are often monotonous and physically demanding, making them ideal for robotic systems equipped with advanced sensors and AI capabilities (Fukui, 2016; Bormann et al., 2019). By delegating such tasks to robots, companies achieve faster order processing and fulfillment while reducing human error. This allows workers to shift their focus to more complex, strategic, and value-added roles, improving job satisfaction and enabling greater operational agility (Chui et al., 2016).

Robotic systems also bring scalability and flexibility to logistics. During periods of peak demand, such as seasonal shopping events or holiday surges, robots can handle increased workloads without the need for additional temporary staff. This ensures orders are processed efficiently and accurately, meeting customer expectations even under high-pressure conditions (Boysen et al., 2019). Furthermore, robotics enhances workplace safety by taking on physically strenuous or hazardous tasks, such as lifting heavy items or operating in extreme conditions. This reduces the risk of workplace injuries and disruptions, creating a safer environment for employees (Adebayo et al., 2024).

In addition to improving efficiency, robotics and automation are driving workforce development by creating opportunities for upskilling. As robotics technologies become more widespread, there is an increasing need for skilled workers who can operate, maintain, and troubleshoot these systems. This shift has led to the emergence of roles such as robotics technicians and automation engineers, offering employees the chance to transition into specialised and higher-paying positions (Brynjolfsson & McAfee, 2014). Companies that invest in training programs not only ensure the smooth integration of robotic systems but also contribute to the professional growth of their employees, fostering a more capable and motivated workforce.

Robotics also promotes inclusivity by enabling individuals with physical disabilities to participate meaningfully in logistics operations. By automating strenuous or repetitive tasks, robotics creates a more accessible work environment, where people of all abilities can contribute effectively to logistics processes (Schlogl & Summer, 2018). This inclusivity aligns with broader goals of diversity and equity, enhancing the reputation of companies that prioritise accessible workplace design.

Overall, the impact of robotics extends beyond individual workplaces, creating ripple effects across the logistics industry and beyond. By increasing efficiency and lowering operational costs, companies can offer faster delivery times and more competitive pricing, enhancing the customer experience, improving resource utilisation and minimising waste through precision operations (Manyika et al., 2017.

- *Wearable Technology:* Complementing robotics, wearable technology is becoming a key enabler of safety and productivity in logistics. Devices such as smart glasses, wristbands, and sensors provide real-time monitoring of workers' health and movements, offering actionable insights to prevent accidents and enhance well-being (Svertoka et al., 2021). For instance, wearable sensors can track metrics such as heart rate, body temperature, and movement patterns, alerting supervisors to potential risks such as fatigue or overheating. Smart glasses enable workers to receive live updates—such as picking instructions or inventory data—directly in their field of vision, improving efficiency and reducing the need for handheld devices (Mladenov et al., 2018).

Wearables also enhance ergonomics and prevent injuries. Wristbands equipped with motion sensors can monitor posture and movements, providing real-time feedback to help workers avoid repetitive strain injuries or improper lifting techniques (Roh et al., 2023). Additionally, wearable devices with location-tracking capabilities improve operational oversight, ensuring that help is available when needed and tasks are completed efficiently (Ding et al., 2020).

Beyond safety, wearable technology fosters a culture of health awareness by empowering workers to take ownership of their well-being. Devices that track activity levels, sleep patterns, and overall health can motivate employees to adopt healthier lifestyles, contributing to long-term improvements in job satisfaction and workplace morale (Perera et al., 2023). By promoting safety, inclusivity, and wellness, wearable technology complements robotics and automation in creating a safer, more productive, and more engaged workforce.

h. Augmented Reality (AR) and Virtual Reality (VR) for Training and Safety

Augmented Reality (AR) and Virtual Reality (VR) are having a significant impact on training and safety in the logistics industry. Traditionally, logistics workers have relied on manuals and static instruction for learning complex tasks. However, AR and VR provide immersive, hands-on experiences that simulate real-world scenarios, allowing workers to practice and develop skills in a risk-free environment (Pinto et al., 2019). This approach enhances the learning process, ensuring that employees are better prepared to perform their tasks efficiently and safely. For example, VR simulations enable forklift operators to practice manoeuvring in tight spaces, a common challenge in busy warehouses. This type of training builds confidence and skill, reducing the likelihood of accidents once they are in real-world situations (Haleem et al., 2020). AR, on the other hand, can overlay digital instructions directly into a worker's field of vision, guiding them step-by-step through complex procedures. This reduces the need for supervisors to intervene constantly and helps workers become more autonomous in their roles (Fitria, 2023).

These technologies also democratise training, making it accessible to individuals in remote or underserved communities. With AR and VR, high-quality training no longer requires proximity to training centers, allowing people from all regions to acquire essential skills and improve their job prospects (Rios and Velasco, 2019). As a result, logistics companies can build a more inclusive workforce while simultaneously improving safety and efficiency.

In addition to training, AR and VR have practical applications in daily operations. AR can provide real-time safety alerts, warning workers of potential hazards in their immediate environment, such as unstable loads or moving machinery. This instant feedback reduces the risk of accidents and creates a safer workplace (Stefan et al., 2023). By integrating AR and VR into their operations, logistics companies are not only improving safety standards but also creating a more engaged, skilled, and confident workforce.

i. Sustainable Energy Solutions and Green Logistics

Despite the heavy carbon footprint of the logistics industry, renewable energy technologies, such as solar-powered warehouses, electric vehicles (EVs), and alternative fuels, are enabling logistics companies to reduce their carbon emissions and align with global sustainability goals (Raza & Lin, 2020). Beyond their environmental benefits, these advancements also promote healthier communities by reducing both air and noise pollution.

- *Solar Energy in Logistics Operations:* Solar energy is a significant driver of sustainable logistics operations and infrastructural development. For instance, warehouses equipped with solar panels generate their own power thereby reducing reliance on fossil fuels and lowering long-term operational costs. Additionally, surplus solar energy can be fed back into the grid, supporting the energy needs of surrounding communities and encouraging broader adoption of renewable energy (Kumar et al., 2023). This dual impact—lowering operational emissions and contributing to the renewable energy ecosystem—demonstrates how logistics companies can extend their positive environmental impact beyond their immediate operations.
- *Electric Vehicles for Greener Logistics:* Electric vehicles (EVs), particularly electric trucks, are envisaged to replace their traditional diesel-powered counterparts. These vehicles produce zero exhaust emissions, significantly improving air quality, particularly in urban areas where congestion amplifies pollution. Their adoption helps mitigate public health risks associated with respiratory illnesses caused by poor air quality, making cities cleaner and healthier (Dablanc et al., 2017; Aryanpur & Rogan, 2024). Furthermore, EVs operate more quietly than diesel trucks, reducing noise pollution and enhancing the quality of life in residential neighbourhoods and urban centers (Jaller & Pahwa, 2020). As advancements in battery technology enable longer ranges, electric trucks are increasingly viable for long-distance deliveries, ensuring sustainability across the supply chain (Aderibigbe & Gumbo, 2023).
- *Alternative Fuel Vehicles:* In addition to EVs, alternative fuels such as biofuels, hydrogen fuel cells, and compressed natural gas (CNG) are gaining traction in the logistics sector. Biofuels, derived from renewable resources like agricultural waste and algae, offer a cleaner alternative to traditional diesel without requiring substantial modifications to existing vehicles (Demirbas,

2018). Hydrogen fuel cells, producing only water vapor as a byproduct, represent one of the cleanest energy solutions available, particularly suited for long-haul deliveries (Matulic et al., 2019). CNG vehicles, while not entirely emissions-free, generate significantly less pollution than diesel trucks and provide a cost-effective bridge solution for companies looking to lower their carbon footprint (Sonnleitner et al., 2021). Together, these alternative fuels are helping logistics companies transition away from fossil fuels, contributing to a more sustainable future.

5.5 Collaborative Logistics and Social Partnerships

In order to successfully embed social value into logistics, collaboration between multiple stakeholders—private companies, governments, and local communities—is crucial. Collaborative logistics leverages the strengths of each participant to build resilient, inclusive, and efficient supply chains that address both business goals and broader social needs (McKinnon, 2018). These partnerships are particularly effective in creating sustainable logistics solutions that enhance social equity by providing employment opportunities, improving access to essential goods and services, and supporting underserved communities.

One major advantage of collaborative logistics is the ability to pool resources and expertise from different sectors. Public-private partnerships (PPPs), for example, allow governments to tap into the technological capabilities and operational efficiency of the private sector, while companies benefit from the support of public policies and infrastructure. By working together, these stakeholders can implement solutions that promote social inclusion, such as providing reliable and affordable transportation to low-income areas or supporting local businesses through more accessible supply chains (Santana Palacios, 2021).

Governments play a pivotal role in establishing the regulatory framework and infrastructure necessary for advancing sustainable logistics practices. A key initiative in this effort is the implementation of low-emission zones in many cities, designed to incentivise companies to adopt cleaner technologies such as electric vehicles. These measures not only reduce urban air pollution but also encourage businesses to align their operations with sustainability goals, fostering a collaborative effort between public and private sectors (Gumzej, 2023). In addition to regulatory measures, governments are offering financial incentives and subsidies to accelerate the adoption of electric and alternative fuel vehicles. These programs lower the initial costs of transitioning to greener technologies, making them more accessible to logistics companies. When combined with the long-term economic benefits—such as reduced fuel consumption and lower maintenance costs—these incentives present a compelling case for investment in sustainable transportation solutions (IEA, 2021). For example, businesses that integrate electric vehicles into their fleets not only decrease their environmental impact but also contribute to the development of cleaner, quieter urban environments, improving the quality of life for local residents.

Private companies, in turn, play an essential role by investing in sustainable energy solutions and smart logistics technologies that enhance efficiency and minimise emissions. Innovations such as solar-powered warehouses, route optimisation software, and alternative fuels complement government initiatives by reducing the logistics sector's overall carbon footprint. By aligning their efforts, governments and businesses can build a cohesive logistics ecosystem that prioritises environmental sustainability while maintaining economic viability.

This partnership between public policy and private enterprise creates a logistics system that benefits both the environment and society. By adopting sustainable practices and leveraging governmental support, logistics companies are not only advancing their competitive edge but also contributing to the health and well-being of communities, particularly in densely populated urban areas. Together, these efforts pave the way for a future where logistics is both efficient and environmentally responsible.

Another important aspect of collaborative logistics is the involvement of local communities. Community partnerships enable logistics providers to better understand the specific needs of the areas they serve, which leads to more effective and inclusive solutions. For instance, engaging with local residents and businesses allows logistics companies to design delivery systems that are sensitive to the unique characteristics of urban and rural settings, such as adjusting delivery times to avoid traffic congestion or implementing micro-distribution hubs in underserved areas (Chung, 2021). These solutions contribute to creating a more equitable distribution of resources and services.

Collaborative efforts also extend to addressing workforce challenges in the logistics sector. The rise of automation and robotics has reshaped the workforce, and collaboration between companies, governments, and educational institutions is key to ensuring workers are equipped with the skills needed for the future. By investing in training programs and creating opportunities for upskilling, these partnerships foster economic inclusion and create a more adaptable workforce. This is especially important in areas where employment opportunities are limited, as logistics companies can provide stable jobs that contribute to local economic growth (Prokopenko et al., 2024).

Furthermore, collaborative logistics initiatives can help address pressing social challenges, such as food security and access to medical supplies. During crises like natural disasters or pandemics, logistics companies and governments must work closely to ensure that essential goods, such as food and medical supplies, reach vulnerable populations in a timely manner. The COVID-19 pandemic highlighted the importance of such collaboration, as logistics networks played a vital role in distributing vaccines and medical equipment globally (Marr, 2020).

5.6 Conclusion

Embedding social value into sustainable transport and smart logistics is essential for addressing global challenges like climate change, urbanisation, and inequality.

This chapter illustrates how technologies such as IoT, AI, blockchain, and robotics can enhance not only operational efficiency but also inclusivity, equity, and resilience in logistics systems. The case study of Bogotá's TransMilenio system exemplifies the transformative potential of integrating social value, demonstrating how collaboration between governments, private enterprises, and communities can lead to equitable and sustainable outcomes. While challenges remain, the system emphasises the importance of partnerships and holistic approaches in creating accessible and environmentally responsible transport networks.

Looking forward, the logistics industry must balance technological advancements with human-centric goals, aligning with frameworks like Industry 5.0 to prioritise societal needs alongside operational objectives. Collaborative efforts across sectors and active community participation are pivotal for creating adaptive systems that foster equity and sustainability. By embedding social value at its core, the logistics sector can evolve into a critical driver of inclusive economic growth, environmental stewardship, and resilient urban development, laying the foundation for a more sustainable future.

Declaration of Generative AI and AI-assisted Technology in the Writing Process

During the preparation of the manuscript, the authors used specialist referencing software Refworks for reference generation and citation; and SCISPACE's typeset.io to discover new papers, and get summary, findings, and conclusions from extant literature. After using these tools, the authors reviewed and edited the content as needed and take full responsibility for the content of the paper.

References

Abirami, S., Pethuraj, M., Uthayakumar, M., & Chitra, P. (2024). A systematic survey on big data and artificial intelligence algorithms for intelligent transportation system. Case Studies on Transport Policy, 17, 101247. https://10.1016/j.cstp.2024.101247.

Adebayo, R.A., Obiuto, N.C., Festus-Ikhuoria, I.C. and Olajiga, O.K. (2024). Robotics in manufacturing: A review of advances in automation and workforce implications. International Journal of Advanced Multidisciplinary Research and Studies, 4(2), 632–638.

Adel, A. and Alani, N.H. (2024). Human-Centric Collaboration and Industry 5.0 Framework in Smart Cities and Communities: Fostering Sustainable Development Goals 3, 4, 9, and 11 in Society 5.0. Smart Cities, 7(4), 1723. https://doi.org/10.3390/smartcities7040068

Aderibigbe, O.O. and Gumbo, T. (2023, September). The role of electric vehicles in greening the environment: Prospects and challenges. In Let It Grow, Let Us Plan, Let It Grow. Nature-based Solutions for Sustainable Resilient Smart Green and Blue Cities. Proceedings of REAL CORP 2023, 28th International Conference on Urban Development, Regional Planning and Information Society (pp. 777–786). CORP–Competence Center of Urban and Regional Planning.

Agrawal, T. K., Angelis, J., Khilji, W.A., Kalaiarasan, R. and Wiktorsson, M. (2023). Demonstration of a blockchain-based framework using smart contracts for supply chain collaboration. International journal of production research, 61(5), 1497–1516.

Aryanpur, V. and Rogan, F. (2024). Decarbonising road freight transport: The role of zero-emission trucks and intangible costs. Scientific Reports, 14(1), 2113.

Autor, D.H. (2015). Why are there still so many jobs? The history and future of workplace automation. Journal of economic perspectives, 29(3), 3–30.

Bajak, M. (2022). Smart Logistics: Sustainable Distribution in the Age of the internet of things. In Sustainable Logistics (pp. 221–245). Productivity Press.

Banister, D. and Hickman, R. (2013). Transport futures: Thinking the unthinkable. Transport Policy, 29, 283-293. https://doi.org/10.1016/j.tranpol.2012.07.005.

Batty, M. (2012). Smart cities, big data. Environment and Planning B: Planning and Design, 39(2), 191–193.

Ben-Daya, M., Hassini, E. and Bahroun, Z. (2019). Internet of things and supply chain management: A literature review. International Journal of Production Research, 57(15-16), 4719–4742.

Biehl, A., Chen, Y., Sanabria-Véaz, K., Uttal, D. and Stathopoulos, A. (2019). Where does active travel fit within local community narratives of mobility space and place?. Transportation research part A: policy and practice, 123, 269–287.

Blasi, S., Ganzaroli, A. and De Noni, I. (2022). Smartening sustainable development in cities: Strengthening the theoretical linkage between smart cities and SDGs. Sustainable Cities and Society, 80, 103793.

Bocarejo, J.P. and Tafur, L.E. (2013). Urban land use transformation driven by an innovative transportation project, Bogotá, Colombia. Global Report on Human Settlements Case Study. UN Habitat. https://unhabitat.org/sites/default/files/2013/06/GRHS.2013.Case_.Study_.Bogota.Colombia.pdf.

Bogue, R. (2016). Growth in e-commerce boosts innovation in the warehouse robot market. Industrial Robot: An International Journal, 43(6), 583–587.

Bormann, R., de Brito, B.F., Lindermayr, J., Omainska, M. and Patel, M. (2019). Towards automated order picking robots for warehouses and retail. In Computer Vision Systems: 12th International Conference, ICVS 2019, Thessaloniki, Greece, September 23–25, 2019, Proceedings 12 (pp. 185–198). Springer International Publishing.

Boysen, N., De Koster, R. and Weidinger, F. (2019). Warehousing in the e-commerce era: A survey. European Journal of Operational Research, 277(2), 396–411.

Brynjolfsson, E., and McAfee, A. (2014). The second machine age: Work, progress, and prosperity in a time of brilliant technologies. WW Norton & company.

Callaghan, V., Chin, J., Doctor, F., Kymäläinen, T., Peña-Rios, A., Phengdy, C. and Zheng, P. (2020). Journeys in the age of smart cities: some fresh perspectives. Handbook of smart cities, 1–39.

Center for Public Impact (2017). TransMilenio: Renewing Bogotá's transport system. Retrieved from https://www.centreforpublicimpact.org/case-study/transmilenio/ [Accessed 25th August 2017].

Chen, W., Men, Y., Fuster, N., Osorio, C. and Juan, A.A. (2024). Artificial Intelligence in Logistics Optimization with Sustainable Criteria: A Review. Sustainability, 16(21), 9145. https://doi.org/10.3390/su16219145.

Choi, K.-S. (2023). Effective demand forecast in supply chain management: Methodology and measure. International Journal of Engineering Research & Technology (IJERT), 12(08), August.

Chui, M., Manyika, J. and Miremadi, M. (2016). Where machines could replace humans-and where they can't (yet). The McKinsey Quarterly, 1–12.

Chung, S.H. (2021). Applications of smart technologies in logistics and transport: A review. Transportation Research Part E: Logistics and Transportation Review, 153, 102455. https://doi.org/10.1016/j.tre.2021.102455.

Colombia Reports (2017). Bogotá socio-economic statistics. Retrieved from https://colombiareports.com/bogota-economy-statistics/ [Accessed 25th August 2017].

Cruz Daraviña, P.A. (2018). Urban freight transport vs sustainable urban planning: Public space renewal, passenger public transport, and land use dynamics. Case study in metropolitan area of Bogotá. Universidad de los Andes. https://repositorio.uniandes.edu.co/entities/publication/fa2f9eb7-4fc1-4605-b1f3-d6690d27fdb7.

Dablanc, L., Morganti, E., Arvidsson, N., Woxenius, J., Browne, M. and Saidi, N. (2017, October). The rise of on-demand 'Instant Deliveries' in European cities. In Supply Chain Forum: An International Journal (Vol. 18, No. 4, pp. 203–217). Taylor & Francis.

Demirbas, A., Bafail, A., Ahmad, W. and Sheikh, M. (2016). Biodiesel production from non-edible plant oils. Energy Exploration & Exploitation, 34(2), 290–318.

Ding, S., Tukker, A. and Ward, H. (2023). Opportunities and risks of internet of things (IoT) technologies for circular business models: A literature review. Journal of environmental management, 336, 117662.

Fitria, T.N. (2023). Augmented reality (AR) and virtual reality (VR) technology in education: Media of teaching and learning: A review. International Journal of Computer and Information System (IJCIS), 4(1), 14–25.

Francisco, K. and Swanson, D. (2018). The supply chain has no clothes: Technology adoption of blockchain for supply chain transparency. Logistics, 2(1), 2.

Fukui, T. (2016, December). A systems approach to big data technology applied to supply chain. In 2016 IEEE International Conference on Big Data (Big Data) (pp. 3732–3736). IEEE.

Garcia-Suarez, C., Rivera-Perez, A.L. and Rodriguez-Valencia, A. (2018). Defining TransMilenio Users' Value and Satisfaction through the Lean Thinking Approach. Transportation Research Record, 2672(8), 455–463.

Garg, A., Maurya, S.P., Laxmi, M.B., Raut, K.J., Yerasuri, S.S. and Shinde, R. M. (2024). Smart Cities and Smart Supply Chain: Integration for Sustainable Urban Development. Power System Technology, 48(1), 239–254.

Gómez Rios, M.D., Paredes Velasco, M. (2019). Augmented Reality as a Methodology to Development of Learning in Programming. In: Botto-Tobar, M., Pizarro, G., Zúñiga-Prieto, M., D'Armas, M., Zúñiga Sánchez, M. (eds) Technology Trends. CITT 2018. Communications in Computer and Information Science, vol 895. Springer, Cham. https://doi.org/10.1007/978-3-030-05532-5_24.

Good Practices in City Energy Efficiency (2017). Bogota, Colombia – Bus Rapid Transit for Urban Transport. Retrieved from https://openknowledge.worldbank.org/bitstream/handle/10986/17533/521260ESMAP0Bo10Box345553B01PUBLIC1.pdf [Accessed 25th August 2017].

Gubbi, J., Buyya, R., Marusic, S. and Palaniswami, M. (2013). Internet of Things (IoT): A vision, architectural elements, and future directions. Future generation computer systems, 29(7), 1645–1660.

Gumzej, R. (2021). Intelligent Logistics Systems for Smart Cities and Communities. Berlin/Heidelberg, Germany: Springer.

Gumzej, R. (2023). Collaborative Logistics in Smart Cities: Integrating Technological Advancements and Social Goals. Journal of Urban Logistics, 12(1), 78–95.

Gunasekaran, A., Subramanian, N. and Papadopoulos, T. (2020). Information technology for competitive advantage within logistics and supply chains: A review. Transportation Research Part E: Logistics and Transportation Review, 138, 101951.

Haleem, A., Javaid, M. and Khan, I.H. (2020). Virtual reality (VR) applications in dentistry: An innovative technology to embrace. Indian Journal of Dental Research, 31(4), 666–667.

Hofmann, E. and Rüsch, M. (2017). Industry 4.0 and the current status as well as future prospects on logistics. Computers in industry, 89, 23–34.

Hopkins, J. and Hawking, P. (2018). Big Data Analytics and IoT in logistics: a case study. The International Journal of Logistics Management, 29(2), 575–591.

Huang, S., Guo, Y., Zha, S. and Wang, Y. (2019). An Internet-of-Things-based production logistics optimisation method for discrete manufacturing. International Journal of Computer Integrated Manufacturing, 32(1), 13–26.

IEA (2021), Global EV Outlook 2021, IEA, Paris https://www.iea.org/reports/global-ev-outlook-2021, Licence: CC BY 4.0

Index Mundi (2017). Colombia Demographics Profile 2017. Retrieved from http://www.indexmundi.com/colombia/demographics_profile.html [Accessed 25th August 2017].

Issaoui, Y., Khiat, A., Bahnasse, A. and Ouajji, H. (2020). Smart Logistics: Blockchain trends and applications. J. Ubiquitous Syst. Pervasive Networks, 12(2), 9–15.

Ivanov, D. and Dolgui, A. (2020). A digital supply chain twin for managing the disruption risks and resilience in the era of Industry 4.0. International Journal of Production Research, 58(23), 7292–7311.

Jaller, M. and Pahwa, A. (2020). Evaluating the environmental impacts of online shopping: A behavioral and transportation approach. Transportation Research Part D: Transport and Environment, 80, 102223.

Jauregui-Fung, F. (2022). Introduction to sustainable public transport solutions in Latin America and the Caribbean (LAC) and Asia. Report for the "Inclusive and sustainable smart cities in the framework of the 2030 Agenda for Sustainable Development" Project.

Kamble, S.S., Gunasekaran, A. and Sharma, R. (2019). Analysis of the driving and dependence power of barriers to adopt industry 4.0 in the Indian manufacturing industry. Journal of Cleaner Production, 213, 977-994.

Kamilaris, A., Fonts, A. and Prenafeta-Boldú, F.X. (2019). The rise of blockchain technology in agriculture and food supply chains. Trends in food science & technology, 91, 640–652.

Kawasaki, T., Namba, Y., Oka, H. and Dulebenets, M.A. (2023). Freight trip distribution using spatiotemporal aggregate data: A modified collective flow diffusion model-based approach. Transportation Research Interdisciplinary Perspectives, 21, 100904.

Klumpp, M. (2018). The index number problem with DEA: Insights from European University efficiency data. Education Sciences, 8(2), 79.

Kshetri, N. (2018). 1 Blockchain's roles in meeting key supply chain management objectives. International Journal of information management, 39, 80–89.

Kumar, C.M.S., Singh, S., Gupta, M.K., Nimdeo, Y.M., Raushan, R., Deorankar, A.V. and Nannaware, A.D. (2023). Solar energy: A promising renewable source for meeting energy demand in Indian agriculture applications. Sustainable Energy Technologies and Assessments, 55, 102905.

Lawrence, J. and Durana, P. (2021). Artificial intelligence-driven big data analytics, predictive maintenance systems, and internet of thingsbased real-time production logistics in sustainable Industry 4.0 wireless networks. Journal of Self-Governance and Management Economics, 9(4), 62–75.

Lee, C.K.M., Lv, Y., Ng, K.K.H., Ho, W. and Choy, K.L. (2018). Design and application of Internet of Things-based warehouse management system for smart logistics. International Journal of Production Research, 56(8), 2753–2768.

Lemoine, P.D., Sarmiento, O.L., Pinzón, J.D. and Meisel, J.D. (2016). TransMilenio, a scalable bus rapid transit system for promoting physical activity. Journal of Urban Health, 93(1), 83-98. https://www.ncbi.nlm.nih.gov/pmc/articles/PMC4835357/.

Li, R., Pereira, F.C. and Ben-Akiva, M.E. (2018). Overview of traffic incident duration analysis and prediction. European transport research review, 10(2), 1–13.

Mackey, T.K. and Nayyar, G. (2017). A review of existing and emerging digital technologies to combat the global trade in fake medicines. Expert opinion on drug safety, 16(5), 587–602.

Manyika, J., et al. (2017) A Future That Works: Automation, Employment and Productivity. McKinsey & Company, New York.

Marr, B. (2020). Coronavirus: How artificial intelligence, data science and technology is used to fight the pandemic. https://www. forbes. com/sites/bernardmarr/2020/0, 3, 13.

Matenga, A.E. and Mpofu, K. (2022). Blockchain-based cloud manufacturing SCM system for collaborative enterprise manufacturing: a case study of transport manufacturing. Applied Sciences, 12(17), 8664.

Matulić, N., Radica, G., Barbir, F. and Nižetić, S. (2019). Commercial vehicle auxiliary loads powered by PEM fuel cell. International journal of hydrogen energy, 44(20), 10082–10090.

McKinnon, A. (2018). Decarbonizing Logistics: Distributing Goods in a Low Carbon World. Kogan Page Publishers.

Mladenov, B., Damiani, L., Giribone, P. and Revetria, R. (2018, October). A short review of the SDKs and wearable devices to be used for an application for an industrial working environment. In Proceedings of the World Congress on Engineering and Computer Science (Vol. 1, pp. 23–25).

Mnyakin, M. (2023). Challenges and Opportunities of Integrating Autonomous Vehicles into Urban Retail Delivery Services. Reviews of Contemporary Business Analytics, 6(1), 14–28.

Murray, C.C. and Chu, A.G. (2015). The flying sidekick travelling salesman problem: Optimisation of drone-assisted parcel delivery. Transportation Research Part C: Emerging Technologies, 54, 86–109.

Nathanail, E., Gogas, M. and Adamos, G. (2016). Smart interconnections of interurban and urban freight transport towards achieving sustainable city logistics. Transportation Research Procedia, 14, 2763-2771. https://doi.org/10.1016/j.trpro.2016.05.078.

Nguyen, T, Zhou, L, Spiegler, V, Ieromonachou, P. and Lin, Y. (2018). Big data analytics in supply chain management: A state-of-the-art literature review, Computers & Operations Research, vol. 98, no. 10, pp. 254- 264. https://doi.org/10.1016/j.cor.2017.07.004.

Nordhoff, S., De Winter, J., Kyriakidis, M., Van Arem, B. and Happee, R. (2018). Acceptance of Driverless Vehicles: Results from a Large Cross-National Questionnaire Study. Journal of advanced transportation, 2018(1), 5382192.

Nurgaliev, I., Eskander, Y. and Lis, K. (2023). The Use of Drones and Autonomous Vehicles in Logistics and Delivery. Logistics and Transport, 57.

Otto, A., Agatz, N., Campbell, J., Golden, B. and Pesch, E. (2018). Optimization approaches for civil applications of unmanned aerial vehicles (UAVs) or aerial drones: A survey. Networks, 72(4), 411–458.

Oviedo D. and Guzman LA. Revisiting Accessibility in a Context of Sustainable Transport: Capabilities and Inequalities in Bogotá. Sustainability. 2020; 12(11):4464. https://doi.org/10.3390/su12114464.

Perdomo Calvo, J.A., Mendoza, C.A., Baquero-Ruiz, A.F. and Mendieta-Lopez, J.C. (2007). Study of the effect of the T ransM ilenio mass transit project on the value of properties in Bogotá, Colombia. Lincoln Institute of Land Policy Working Paper No. WP07CA1.

Pereira, J., Souza, R., Oliveira, J. and Moita, A. (2024). Hydrogen Production, Transporting and Storage Processes—A Brief Review. Clean Technologies, 6(3), 1260–1313.

Perera, M.S.D., Jayathilake, S.R., Ranasinghe, J.D., Bartholomeusz, S.V., Chathuranga, H.S., Vidhanaarachchi, S. and Dasanayaka, A. (2023). Stress Monitoring and Relieving Application for IT Professionals. International Research Journal of Innovations in Engineering and Technology, 7(10), 609.

Pinto, A.S., Abreu, A., Costa, E. and Paiva, J. (2022). Augmented reality for a new reality: using UTAUT-3 to assess the adoption of mobile augmented reality in tourism (MART). Journal of Information Systems Engineering and Management, 7(2), 14550.

Pocacito (2017). Transport and mobility - Bus Rapid Transit: TransMilenio. Retrieved from http://pocacito.eu/sites/default/files/TransMilenio_Bogota.pdf [Accessed 25th August 2017].

PPP IRC (2017). TransMilenio Phase I PPP Contract for Bus Transport System. Retrieved from https://ppp.worldbank.org/public-private-partnership/library/transmilenio-phase-i-ppp-contract-bus-transport-system [Accessed 25th August 2017].

PRI (2015). 8 million people. No subway. Can this city thrive without one?. Retrieved from https://www.pri.org/stories/2015-10-21/can-modern-megacity-bogot-get-without-subway [Accessed 25th August 2017].

Prokopenko, O., Prause, G., Bielialov, T., Jarvis, M., Holovanenko, M. and Kara, I. (2024). Sustainable logistics and passenger transport in smart cities. Acta Logistica (AL), 11(1). https://doi.org/10.22306/al.v11i1.448.

Queiroz, M.M., Telles, R. and Bonilla, S.H. (2019). Blockchain and supply chain management integration: A systematic review of the literature. International Journal of Production Research, 57(7), 2035–2053.

Raza, M.Y. and Lin, B. (2020). Decoupling and mitigation potential analysis of CO2 emissions from Pakistan's transport sector. Science of the Total Environment, 730, 139000.

Roh, T., Esomonu, C., Hendricks, J., Aggarwal, A., Hasan, N.T. and Benden, M. (2023). Examining workweek variations in computer usage patterns: An application of ergonomic monitoring software. Plos one, 18(7), e0287976.

Saberi, S., Kouhizadeh, M., Sarkis, J. and Shen, L. (2019). Blockchain technology and its relationships to sustainable supply chain management. International journal of production research, 57(7), 2117–2135.

Santana Palacios, D. (2021). The Evolution of Bus Rapid Transit (BRT) Systems: Case Studies in Latin America. Sustainable Transportation Journal, 25(3), 105–120.

Santana Palacios, M. (2021). Public transit innovations, equity, and blind spots: Examining the benefits and unintended consequences of investments in public transport projects in Bogotá. eScholarship, University of California. https://escholarship.org/uc/item/5ns8s78k.

Schlogl, L. and Sumner, A. (2018). The rise of the robot reserve army: Automation and the future of economic development, work, and wages in developing countries. SSRN.

Sharma, V., Raut, R.D., Govindarajan, U.H. and Narkhede, B. E. (2022). Advancements in urban logistics toward smart, sustainable reforms in developing enabling technologies and markets. Kybernetes, 51(3), 1038–1061.

Shee, H.K., Miah, S.J. and De Vass, T. (2021), "Impact of smart logistics on smart city sustainable performance: an empirical investigation", The International Journal of Logistics Management, Vol. 32 No. 3, pp. 821-845. https://doi.org/10.1108/IJLM-07-2020–0282.

Sodiya, E.O., Umoga, U.J., Amoo, O.O. and Atadoga, A. (2024). AI-driven warehouse automation: A comprehensive review of systems. GSC Advanced Research and Reviews, 18(2), 272–282.

Sonnleitner, A., Bacovsky, D., Zhang, S., Wu, Y., Nylund, N.O., Kobayashi, M. and Wang, M. (2021). Lessons Learned from Alternative Fuels Experience.

Stefan, H., Mortimer, M. and Horan, B. (2023). Evaluating the effectiveness of virtual reality for safety-relevant training: a systematic review. Virtual Reality, 27(4), 2839–2869.

Svertoka, E., Saafi, S., Rusu-Casandra, A., Burget, R., Marghescu, I., Hosek, J. and Ometov, A. (2021). Wearables for industrial work safety: A survey. Sensors, 21(11), 3844.

Tao, X., Cheng, L., Zhang, R., Chan, W.K., Chao, H. and Qin, J. (2023). Towards Green Innovation in Smart Cities: Leveraging Traffic Flow Prediction with Machine Learning Algorithms for Sustainable Transportation Systems. Sustainability, 16(1), 251.

Toorajipour, R., Sohrabpour, V., Nazarpour, A., Oghazi, P. and Fischl, M. (2021). Artificial intelligence in supply chain management: A systematic literature review. Journal of Business Research, 122, 502–517.

Treiblmaier, H. (2019). Combining blockchain technology and the physical internet to achieve triple bottom line sustainability: a comprehensive research agenda for modern logistics and supply chain management. Logistics, 3(1), 10.

Visan, M., Negrea, S.L. and Mone, F. (2022). Towards intelligent public transport systems in Smart Cities; Collaborative decisions to be made. Procedia Computer Science, 199, 1221–1228.

Wang, Y., Singgih, M., Wang, J. and Rit, M. (2020). Making sense of blockchain technology: How will it transform supply chains? Transportation Research Part E: Logistics and Transportation Review, 136, 101886.

Wu, L., Yue, X., Jin, A. and Yen, D.C. (2016). Smart supply chain management: a review and implications for future research. The international journal of logistics management, 27(2), 395–417.

6 Extended Reality in Warehousing: Bridging Industry 4.0 and 5.0 for Smart Logistics

Mohammadamin Hosseinifard[1*]

6.1 Introduction

The advent of Industry 4.0 has transformed industrial operations by integrating cutting-edge technologies such as the Internet of Things (IoT), Extended Realities (XR), and advanced analytics to enhance interconnectivity between systems. This transformation has particularly revolutionized warehousing, where Industry 4.0 has leveraged XR—comprising Augmented Reality (AR), Virtual Reality (VR), and Mixed Reality (MR)—to optimize critical functions like inventory management, human navigation, and order picking and confirmation. These applications have significantly improved accuracy, speed, and operational reliability, setting a new standard for smart warehousing. By providing real-time data and immersive solutions, these technologies enable companies to adapt to evolving market demands and maintain operational excellence (Perotti et al., 2022).

On the other side, Industry 5.0 represents a human-centric evolution of industrial practices. This paradigm emphasizes the harmonious integration of advanced systems with human workers, prioritizing safety, sustainability, and collaboration. Unlike Industry 4.0, which focuses on automation and interconnectivity, Industry 5.0 aims to create environments that support human roles rather than replace them. By integrating ethical AI, adaptive automation, and worker-focused enhancements, Industry 5.0 ensures that technological innovation aligns with human values, fostering sustainable practices and empowering workers within collaborative frameworks. In warehousing, these principles translate to worker-centric designs, AI-driven safety protocols, and adaptive XR solutions that enhance worker well-being while maintaining high operational standards (Grosse et al., 2023).

Extended Reality bridges the gap between these two paradigms by providing a versatile technological toolkit that aligns with the efficiency goals of Industry 4.0 and the human-centric priorities of Industry 5.0. XR technologies—comprising AR, VR, and MR—play a transformative role in warehousing by enhancing worker interactions with digital systems and physical environments. AR overlays digital information onto the physical environment, enabling workers to visualize

[1] Department of Civil and Industrial Engineering, University of Pisa, Pisa, Italy.
* Corresponding author: Mohammad.Hosseinifard@phd.unipi.it

DOI: 10.1201/9781003466451-6

inventory, navigate warehouses, and streamline order picking with precision. For example, AR-based navigation systems can reduce the time spent to locate items by providing direct visual cues, enhancing workflow efficiency. VR creates fully immersive environments that are ideal for employee training, allowing workers to safely simulate and practice complex tasks in a controlled virtual space. This reduces onboarding time and minimizes errors, as employees are better prepared for real-world challenges. MR blends the physical and digital worlds, enabling real-time interaction between workers, robots, and digital overlays for advanced Human-Robot Collaboration (HRC). MR facilitates tasks like collaborative item picking and maintenance operations, where robots and humans work together seamlessly. By leveraging MR-powered tools such as Microsoft HoloLens, warehouses can improve accuracy and productivity in high-stakes operations (Kurschl et al., 2021); (Grabowski and Wodzyński, 2022).

These XR applications not only streamline operations but also promote a safer, more adaptable, and human-friendly workplace. However, the successful adoption of XR faces several challenges, e.g., high initial costs, workforce adaptation, and technological infrastructure limitations. Addressing these barriers requires strategic implementation plans, collaboration with technology providers, and robust training initiatives. Furthermore, data security and privacy concerns must be prioritized to ensure long-term sustainability (Speicher et al., 2019); (Akbari et al., 2022). As XR technologies continue to evolve, their integration into warehousing systems becomes increasingly vital for maintaining competitiveness and operational resilience.

By connecting the technological advances of Industry 4.0 with the human-centered ethos of Industry 5.0, XR technologies establish a unified framework that drives operational excellence, fosters innovation, and ensures sustainable practices in warehousing. This evolution underscores the importance of aligning technology with human values, paving the way for smarter, more inclusive industrial systems. In the following sections, this chapter will explore how XR technologies can be effectively integrated into warehousing, the potential benefits and challenges of implementation, and the broader implications of adopting Industry 5.0 principles in logistics and supply chain management.

6.2 XR in Warehousing: A Literature Review of Innovations

XR technologies have emerged as a cornerstone of the Industry 4.0 revolution, driving transformative innovations in the warehousing sector by enhancing efficiency, modernizing workflows, and fostering smarter operations. These technologies have enhanced operational efficiency, workforce training, and overall warehouse management processes. One of the most notable advancements is the modernization of traditional manual picking systems. Human operators are now equipped with AR tools, allowing for greater precision, reduced error rates, and improved productivity. These innovations have reshaped warehousing workflows, creating smarter and data-driven environments.

However, the rapid pace of technological development, coupled with evolving industrial scenarios and regulatory frameworks, has accelerated the transition toward Industry 5.0. This next-generation industrial paradigm emphasizes enhanced human-machine collaboration. Robotic systems are increasingly taking on roles such as automated item picking and packing, while XR technologies continue to improve human-robot interfaces. Recent advancements in Artificial Intelligence (AI) have further enabled these interfaces, creating seamless and efficient interactions between humans and machines.

In this section, readers will learn about the key milestones in XR's evolution, from its role as an innovation-driven tool in Industry 4.0 to its current function as a critical enabler of human-centric automation in Industry 5.0. Figure 6.1 provides a visual representation of this progression, which will be discussed in detail in the following sections, along with its significant impact on modern warehousing.

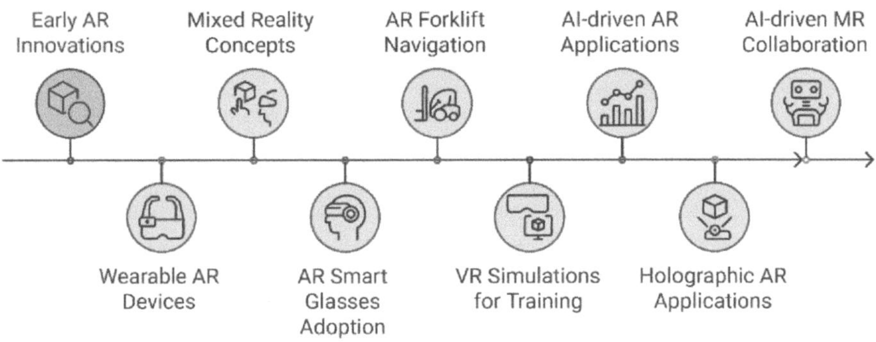

Fig. 6.1 XR milestone in warehouse development (*Source:* The author's elaboration).

The journey began in 2015, with experimental applications of AR in order-picking systems. These early studies demonstrated the potential of AR headsets to reduce picking errors and enhance worker efficiency by overlaying item locations and instructions from the worker's field of view. By 2017, logistics companies began piloting wearable AR devices like Google Glass to track inventory and streamline workflows. These pilot projects revealed both the promise and limitations of early AR hardware, such as limited battery life and restricted fields of view (Blanco-Novoa et al., 2018). The introduction of MR in 2018 marked a significant leap forward. Unlike AR, which overlays digital elements in the physical world, MR allows users to interact with these elements in real time, enabling dynamic decision-making in warehouse environments (Speicher et al., 2019). By 2019, XR technologies moved beyond experimental stages to broader adoption. AR smart glasses became standard tools in logistics operations, enabling hands-free inventory management and route optimization (Plakas et al., 2020). In the following year, AR-assisted forklift navigation systems were introduced, enhancing safety by providing real-time alerts and navigation guidance (Piardi et al., 2019). The focus shifted toward workforce empowerment in 2020, with VR simulations becoming a transformative training tool in warehouse environments. These immersive tools allowed employees to practice handling complex tasks such as forklift operation, inventory management, and hazardous scenario navigation in a risk-free virtual space. VR training systems replicated real-world challenges, enabling users to engage in realistic, interactive simulations that improved skill acquisition and error reduction. For example, studies demonstrated how VR-based order picking and warehouse management systems enhanced user learning and operational efficiency by integrating real inventory data and workflows into the training environment (Yigitbas et al., 2020). The integration of AI-powered systems with advanced technologies in 2022 significantly advanced warehouse inventory management. AI-driven predictive analytics enabled warehouses to forecast inventory shortages, streamline restocking processes, and enhance allocation strategies. These systems provided actionable insights by analyzing real-time data, reducing downtime and waste while improving operational efficiency. This marked a pivotal shift in inventory management practices, showcasing the transformative synergy between AI and automation technologies in optimizing resource utilization and ensuring seamless warehouse operations (Singh and Adhikari, 2023). In 2023, real-time MR and AI technologies were integrated for dynamic inventory management and warehouse layout optimization. Using MR devices such as the Microsoft HoloLens, warehouses adopted systems that enabled operators to interact with real-time digital twins, facilitating object recognition and spatial adjustments. This integration enhanced operational efficiency by optimizing inventory placement, improving retrieval accuracy, and reducing response times in logistics operations (Tang et al., 2023).

The future holds even greater promise. By 2025, AI-driven MR systems will enable seamless collaboration between humans and robots for item picking and sorting. These systems will integrate predictive analytics and contextual awareness, creating warehouses that are not only efficient but also adaptive to dynamic demands.

6.3 Overview of XR Adoption in Smart Warehousing

Smart warehousing has become a cornerstone of modern supply chain management, driven by the need for innovation and efficiency in increasingly complex operational environments. Extended Reality technologies have emerged as a key enabler of this transformation, offering solutions that enhance performance, precision, and adaptability across warehousing processes.

As depicted in Figure 6.2, XR applications are diverse, targeting critical areas such as inventory management, order fulfillment, workforce training, and operational planning. AR supports real-time visualization of data, enabling seamless interaction between human operators and robotic systems while improving tasks such as picking and sorting. VR, on the other hand, is instrumental in creating immersive training simulations, providing a safe and effective platform for skill development and warehouse layout optimization. MR bridges the physical and digital worlds, facilitating on-site control, IoT integration, and data-driven decision-making in real time.

This section provides a foundation for exploring specific XR applications in warehousing, demonstrating how these technologies are fostering a smarter, more connected, and responsive operational environment.

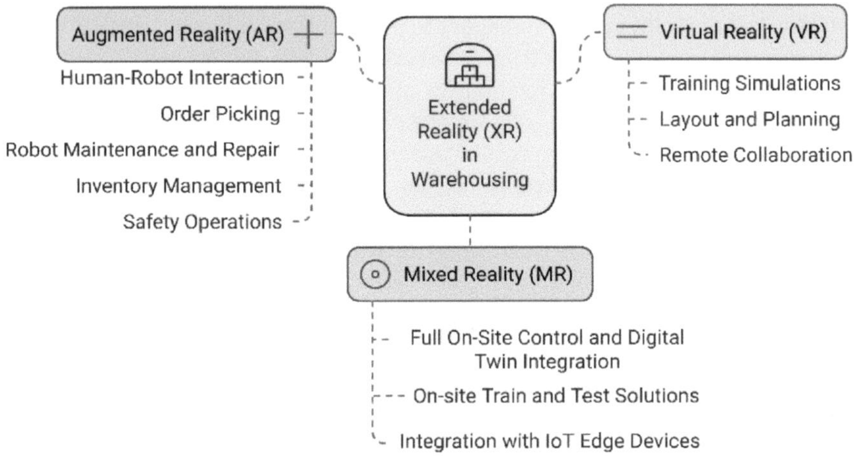

Fig. 6.2 XR applications in smart warehousing: enhancing efficiency and collaboration in Industry 4.0 and 5.0 (*Source:* The author's elaboration).

6.3.1 AR Applications in Warehousing

Enhancing Human-Robot Collaboration

AR serves as a critical interface for human-robot collaboration, fostering seamless interactions between warehouse operators and robotic systems. AR devices, e.g., head-mounted displays (HMDs), overlay real-time instructions and data onto a worker's field of view, improving task accuracy and efficiency. This technology assists workers in executing complex tasks, i.e., picking, packing, and cubing, with enhanced precision and reduced errors (Limeira et al., 2020). In dynamic warehouse settings, AR optimizes navigation and workflow by providing visual path-sharing between robots and human workers, reducing the risk of collisions in chaotic storage environments (Papcun et al., 2019). By aligning AR tools with Industry 5.0 goals, warehousing operations can achieve greater collaboration, safety, and operational precision.

Improving Order-Picking Efficiency

Order picking is one of the most labor-intensive tasks in warehousing. AR-enabled systems like pick-by-vision solutions offer real-time visual guidance, displaying item locations and quantities directly within a worker's field of view. This visual assistance reduces cognitive load and enables faster, more accurate picking (Nagda et al., 2019). Technologies such as RASPICK further integrate AR with Warehouse Management Systems (WMS), ensuring real-time inventory updates and route optimization. Research indicates that AR-assisted picking significantly reduces errors while improving worker productivity (Jumahat et al., 2022); (Matsumoto et al., 2019). These advancements align AR's role with modern logistical demands, enhancing both accuracy and speed in order fulfillment.

AR for Maintenance and Safety Operations

AR is increasingly applied to robot maintenance and repair by providing technicians with step-by-step visual instructions overlaid with physical components. Devices such as smart glasses enable workers to diagnose and repair machinery efficiently, minimizing downtime (Papcun et al., 2019). For safety operations, AR can project hazard warnings, spatial restrictions, and dynamic evacuation routes in real time, significantly improving situational awareness and reducing workplace accidents (Akhmetov and Varol, 2023).

6.3.2 VR Applications in Warehousing

Advanced Worker Training

Virtual Reality provides immersive and risk-free training simulations, enabling warehouse workers to practice complex tasks, e.g., order picking, equipment handling, and emergency responses. By simulating real-world warehouse scenarios, VR improves worker preparedness while reducing the costs and logistical challenges of on-site training (Yigitbas et al., 2020). VR systems replicate warehouse environments and workflows, ensuring that trainees gain hands-on experience. These systems have been shown to increase worker confidence and performance consistency, making VR an essential tool for workforce development in modern warehousing.

Warehouse Layout and Planning

VR supports the design and optimization of warehouse layouts by providing immersive visualization tools. Warehouse managers can test and refine warehouse configurations within virtual environments, which helps identify bottlenecks, safety hazards, and ergonomic inefficiencies before implementation (Havard et al., 2019). Collaborative VR tools further enable multi-user interaction, allowing teams to refine designs dynamically with real-time performance metrics and KPIs.

Networked Warehouse Management

VR offers centralized control through real-time virtual dashboards for large-scale and networked operations. Systems like WareVR integrate drone feeds and IoT data into immersive environments, enabling remote monitoring of inventory, operations, and safety compliance (Kalinov et al., 2021). This integration enhances warehouse visibility, allowing operators to identify inefficiencies, monitor equipment, and predict maintenance needs before disruptions occur.

The potential of VR to simulate 3D digital twins marks a significant evolution in networked warehouse management. Digital twins serve as virtual mirrors of physical warehouses, integrating real-time data from IoT devices, robotics, and inventory systems. These virtual models empower warehouse managers to experiment with different scenarios, refine workflows, and adapt swiftly to shifts in demand or operational challenges. For example, managers might use these simulations to streamline order processing routes, reduce bottlenecks, or examine energy consumption in order to enhance sustainability practices. With continuous technological enhancements, VR is poised to further enrich large-scale warehouse networks by providing predictive analytics, automatic updates, and comprehensive immersive views of operations. This capability significantly bolsters decision-making, efficiency, and scalability within contemporary warehousing environments.

6.3.3 MR Applications in Warehousing

Bridging On-Site and Remote Operations

MR combines the immersive capabilities of VR with AR's real-time data overlays, enabling both on-site and remote warehouse operations. For instance, on-site workers equipped with MR devices can view inventory data, machine diagnostics, or hazard warnings overlaid onto physical objects, while remote supervisors monitor operations virtually in real time. This dual functionality ensures seamless collaboration and improves operational flow.

Enhancing Human-Robot Collaboration

MR improves coordination between workers and robots by displaying live navigation paths, operational statuses, and safety zones. Workers can anticipate robot movements and adjust workflows, reducing risks and improving productivity. For example, MR tools overlay dynamic visual cues to guide workers through Automated Guided Vehicle (AGV) interactions, enhancing safety and task precision.

Optimizing Training and Testing

MR offers an innovative approach to on-site training by blending real-world tasks with virtual guidance. Workers can practice equipment handling or maintenance tasks with augmented instructions overlaid onto physical machinery, enhancing learning outcomes. This method ensures faster skill acquisition and improves operational preparedness, particularly for complex tasks (Gonzalez-Franco et al., 2017).

IoT-Integrated Warehouse Operations

The integration of MR with IoT technologies further enhances inventory management, safety, and real-time decision-making. MR devices visualize IoT sensor data such as temperature, humidity, and stock levels directly on shelves or equipment, enabling immediate corrective actions. In emergency situations, MR guides workers through optimized safety procedures, minimizing risks and ensuring compliance.

Conclusion of XR Applications in Smart Warehousing

Extended Reality technologies, including AR, VR, and MR, are transforming warehousing by enhancing efficiency, collaboration, and decision-making. AR optimizes real-time operations and human-robot interaction, VR revolutionizes training and warehouse design, and MR bridges digital and physical workflows for smarter management. While these technologies align warehousing with Industry 5.0, challenges like costs and workforce adaptation remain. The next section explores these barriers and potential solutions for successful XR implementation.

6.4 Challenges in Implementing XR in Warehousing

Implementing XR technologies in warehousing presents a range of challenges, including technical, financial, and human factors. Overcoming these obstacles is essential to leverage XR's potential to enhance logistics efficiency and accuracy. These challenges often stem from limitations in hardware and software, significant investment costs, and workforce adaptation needs. Moreover, organizations face concerns over data security and regulatory compliance. Safeguarding sensitive data is paramount, especially as XR systems increasingly interact with internal workflows. Navigating compliance with industrial standards and data privacy regulations further complicates adoption, demanding comprehensive planning and robust technological solutions.

6.4.1 Technical Challenges of XR Implementation in Warehousing

Deploying XR in warehousing involves addressing specific technical limitations that affect device usability and system efficiency. These challenges include constraints in device design, such as Field of View (FoV) and weight, which directly influence operational performance as highlighted in Table 1.

FoV and device weight are critical considerations in XR usability. A narrow FoV, such as the 52° offered by the Microsoft HoloLens 2, restricts peripheral awareness, which is essential for spatially intensive tasks in warehouses. Similarly, devices like the HTC Vive Focus 3, weighing 785 grams, can lead to user fatigue during prolonged use, reducing worker productivity. High resolutions and refresh rates, such as the Meta Quest 3's adjustable refresh rate of up to 120 Hz, enhance visual fidelity but place significant demands on processing power and battery capacity. Frequent recharging interrupts operational continuity, creating inefficiencies in fast-paced environments. These ergonomic and performance-related challenges highlight the need for advancements in device design and energy-efficient hardware.

Seamless integration of XR systems with WMS and other automation systems is another critical factor. Compatibility issues and unreliable wireless connectivity can lead to delays or errors, undermining the potential benefits of XR technologies. Real-time data exchange and interoperability are essential to maintain operational consistency and support complex warehouse tasks effectively.

Addressing these challenges requires a coordinated effort to improve hardware specifications, optimize system compatibility, and enhance energy efficiency. As XR technologies evolve, focusing on expanded FoV, reduced device weight, and reliable integration with existing systems will be pivotal in enabling their practical application in logistics.

6.4.2 Cost Barriers to XR Implementation in Warehousing

High costs pose a significant barrier to implementing XR technologies in warehousing, covering everything from the initial purchase of devices to ongoing maintenance and training expenses. The initial investment in XR devices, as detailed in Table 1, includes a range of prices for head-mounted displays such as AR smart glasses and VR headsets, which can cost anywhere from \$500 to over \$3,000 per unit. High-end MR devices like the Microsoft HoloLens or Magic Leap systems are particularly expensive, with prices exceeding \$3,500. In addition to the hardware costs, warehouses must consider expenses associated with custom software development, integration with WMS, and ongoing updates to ensure compatibility with rapidly advancing technologies. These costs can pose a significant challenge, especially for small to medium-sized enterprises (SMEs) that operate on tight budgets and may struggle to justify such investments without immediate, tangible returns.

Beyond the initial investment, maintenance, training, and infrastructure costs further compound the financial burden of XR adoption. For instance, deploying XR in a warehouse requires robust connectivity infrastructure, i.e., high-speed Wi-Fi or 5G networks, to support real-time data synchronization and device functionality. Training employees to effectively use XR devices also adds to operational costs, requiring tailored programs to familiarize workers with the interface and features of these technologies. Furthermore, the need for specialized technical support to maintain XR systems can escalate expenses over time. Although XR technologies have the potential to streamline operations, reduce errors, and enhance productivity, the upfront and ongoing costs serve as a substantial barrier to adoption for many warehouses, delaying the broader implementation of these transformative tools.

Table 6.1 Comparative specifications of recent XR headsets (Source: The author's elaboration)

XR Headset	*Type*	*Resolution (per eye)*	*Field of View*	*Refresh Rate*	*DoF*	*Software Platform*	*Weight*	*Price*
Varjo XR-4	VR/MR	3840 x 3744	120° x 105°	90 Hz	6DoF	Varjo Base	665 g	Starting at $3,990
HTC Vive Focus 3	VR/MR	2448 x 2448	120°	90 Hz	6DoF	Android-based	785 g	$1,300
HTC VIVE XR Elite	VR/MR	2448 x 2448	110°	90 Hz	6DoF	Android-based	625 g	$999
Meta Quest 3	VR/MR	2064 x 2208	110°	72-120 Hz	6DoF	Android-based	515 g	Starting at $499.99
Microsoft HoloLens 2	AR	2k 3:2 light engines	52°	120 Hz	6DoF	Windows Holographic OS	566 g	Starting at $3,500

6.4.3 User Acceptance and Training Challenges

User acceptance and training pose significant challenges in adopting XR technologies in warehousing, as workers often view these new tools as complex and potentially disruptive. This resistance frequently stems from a lack of familiarity with technology and concerns about its potential to disrupt established workflows. To address these apprehensions, organizations must emphasize XR's role as a tool to enhance, rather than replace, human labor.

DHL's Vision Picking program, first tested in 2014 at the company's Dutch facility in collaboration with Google, Vuzix, and Ubimax, provides a strong example of how AR can improve warehouse operations. The system operates on an Ubimax platform, where workers use Google Glass or Vuzix smart glasses along with a ring scanner to facilitate order picking. As shown in Figure 6.3, the AR glasses display digital picking lists, including aisle numbers, exact item locations, required quantities, and sorting guidance. This hands-free system improves efficiency by allowing workers to quickly identify, pick, and place items in designated boxes on a cart while minimizing errors. Additionally, it reduces training time and helps bridge language barriers for migrant workers. DHL's trials demonstrated a 40% reduction in picking errors and faster orientation for new employees, underscoring AR's potential to enhance logistics operations (Glockner et al., 2014).

Training programs designed to facilitate XR adoption should bridge the knowledge gap and build worker confidence. A phased approach can include introductory tutorials, scenario-based training, and advanced simulations. For instance, DHL's approach reduced the reliance on lengthy manuals and increased hands-on interaction with AR tools, enabling workers to adapt quickly to new technologies. These structured programs provide an intuitive learning curve that ensures competency while minimizing apprehension.

Ergonomic designs and user-friendly interfaces are critical to fostering acceptance. Lightweight devices and intuitive software reduce physical and cognitive strain, making XR tools more accessible for extended use. DHL's AR systems, which incorporate features like hands-free navigation and automated barcode scanning, highlight the importance of usability in high-pressure environments.

Effective user training is crucial for increasing the acceptance rate of XR applications in the industry. Training programs should focus on building worker familiarity with XR devices and applications, ensuring users understand how to operate and integrate these tools into their workflows. By offering clear, hands-on instruction and addressing common concerns through structured sessions, organizations can foster confidence and competence, ultimately enhancing both worker satisfaction and operational efficiency.

Ultimately, overcoming resistance to XR adoption requires a balance between technological innovation and human-centric design. Collaborative efforts among management, developers, and workers are essential. By investing in ergonomic systems, comprehensive training, and a culture of openness to change,

Fig. 6.3 DHL Vision Picking System – AR-enabled smart glasses provide real-time picking instructions, improving efficiency, accuracy, and hands-free operation in warehouse logistics - (Adapted from an open-source paper) (Husár and Knapčíková, 2021).

organizations can unlock XR's potential to enhance warehouse efficiency while addressing workforce concerns. These strategies are fundamental to achieving the transformative benefits XR technologies offer in industrial applications.

6.4.4 Data Security and Privacy Challenges in XR

XR technologies transform warehouse operations but simultaneously expose critical data to new security and privacy risks, stemming from their extensive reliance on interconnected and data-intensive systems. As highlighted by de Guzman et al. (2019), XR systems inherently rely on vast volumes of sensitive data—capturing everything from spatial layouts and inventory levels to worker movements and real-time interactions. This extensive data ecosystem introduces vulnerabilities that, if exploited, could compromise the confidentiality, integrity, or availability of critical warehouse information. Ensuring robust security measures is essential but remains a substantial barrier to XR adoption (De Guzman et al., 2019).

A primary concern is the risk of unauthorized access to sensitive data collected by XR devices. In warehouses, AR smart glasses or MR headsets continuously gather real-time data to overlay digital information onto the physical environment. This data includes not only operational details, e.g., inventory and logistics routes, but also personal information about employees, including their identities and task histories. Without appropriate safeguards like encryption and role-based access controls, this information can become a target for cyberattacks, potentially exposing warehouses to operational disruptions and privacy breaches.

Furthermore, the integration of XR systems with WMS and cloud-based platforms amplifies data security risks. These interconnected systems rely on high-speed data transmission to ensure real-time performance, yet they are susceptible to threats like data interception, unauthorized data manipulation, and system breaches. Techniques such as end-to-end encryption and anonymization are often employed to mitigate these risks, but they can impose trade-offs on system responsiveness—a critical requirement in fast-paced warehouse environments. Ensuring both security and performance requires significant technical expertise and investment, which can deter organizations from fully embracing XR technologies.

Collaboration through XR technologies adds another layer of complexity to data security and privacy. In networked warehouses or remote monitoring scenarios, multiple users might interact with shared virtual environments, increasing the potential for unauthorized data access or manipulation. To counter this, robust authentication systems, user activity tracking, and secure communication protocols are necessary. However, implementing these measures in cost-sensitive logistics environments is challenging, as it requires both technical sophistication and substantial financial investment, further complicating XR adoption.

To overcome these barriers, warehouses must adopt a holistic approach to security, combining advanced technological solutions with comprehensive training and policy frameworks. By addressing data security and privacy concerns proactively, organizations can unlock the potential of XR technologies while safeguarding sensitive information, ensuring smoother adoption and integration into warehouse operations. The findings of de Guzman et al. (2019) underscore the importance of developing secure and user-centric XR ecosystems, particularly in data-intensive industries like logistics and warehousing.

6.4.5 Regulatory and Compliance Issues in Implementing XR Solutions

The integration of XR technologies into warehouse operations is shaped by a dense web of regulatory and compliance demands, requiring meticulous alignment with industry standards and legal frameworks to ensure seamless and secure implementation. A critical regulatory challenge lies in securing access and licenses for integrating XR systems with WMS and industrial automation platforms. These systems often operate under proprietary architectures, with strict protocols governing how external interfaces, such as XR applications, can access their data. Organizations implementing XR must ensure that their solutions meet these protocols and obtain necessary licenses, which can add significant complexity and cost to the deployment process.

Regulations governing industrial automation systems, particularly those linked to IoT-enabled WMS platforms, further complicate XR adoption. Such regulations often mandate stringent cybersecurity measures to protect the integrity of real-time operational data. XR interfaces that link with IoT sensors, robots, and inventory systems must comply with industrial standards to ensure secure and reliable data exchange. For example, XR solutions must support encrypted communication and adhere to frameworks like the NIST Cybersecurity Framework or ISO/IEC 27001 for information security management. These regulatory requirements are critical to preventing unauthorized access to sensitive data but can restrict the ease of integration, demanding advanced technical expertise and compliance-focused development practices.

Another key aspect of compliance involves managing restrictions around data privacy and access. XR systems that interact with WMS and IoT platforms must handle sensitive operational data, such as inventory statuses, robotic paths, and employee activity logs. Regulations such as the General Data Protection Regulation (GDPR) in Europe or sector-specific standards like the Federal Information Security Management Act (FISMA) in the U.S. impose strict requirements on how this data is collected, stored, and shared. XR applications must incorporate robust access controls to ensure that only authorized personnel can retrieve or manipulate this information. These measures, while necessary for maintaining privacy and data integrity, can limit the operational flexibility of XR solutions and pose a barrier to widespread implementation.

Cybersecurity concerns further exacerbate regulatory challenges, as XR systems introduce new potential vulnerabilities to WMS and IoT-linked platforms. The ability of XR applications to interface with multiple systems increases the potential attack surface for cyber threats, prompting regulators to impose stricter controls on integration processes. For example, ensuring that XR devices do not act as weak links in industrial networks requires extensive testing, adherence to security certifications, and ongoing monitoring. These measures not only increase the cost and complexity of XR deployment but also slow down the implementation timeline, making compliance a significant hurdle for organizations adopting XR in warehousing environments. Addressing these challenges requires a collaborative effort among XR developers, system integrators, and regulatory bodies to streamline standards and establish trust in XR solutions for industrial applications.

6.5 Future of XR in Warehousing

6.5.1 Future Trends in AI Integration with XR for Warehousing

The integration of AI with XR in warehousing is increasingly focused on enhancing image processing capabilities in item picking. AI-driven computer vision systems are being developed to accurately identify, sort, and manage inventory by automating visual tasks that were previously manual and error-prone. These systems leverage deep learning algorithms to improve the accuracy and speed of object recognition, even under varying lighting and environmental conditions. This trend not only boosts operational efficiency but also significantly reduces the occurrence of errors in inventory management, thereby optimizing the workflow and resource allocation within warehouses.

Another pivotal area of development is the use of AI to enhance human-robot interaction interfaces. As warehouses increasingly adopt robotic solutions to handle repetitive and physically demanding tasks, the need for sophisticated interfaces that allow seamless human-robot collaboration becomes critical. AI technologies are being utilized to develop intuitive communication systems where robots can understand and predict human actions and vice versa. This involves the use of sensors and AI algorithms that help robots interpret human gestures and commands more effectively, enhancing collaborative tasks and safety in shared workspaces.

AI is also transforming training protocols within warehousing through immersive VR and MR environments. By simulating real-world scenarios, AI-enhanced XR training platforms can offer personalized learning experiences that adapt to the user's performance and learning pace. These platforms are capable of generating dynamic content that reacts to user interactions, providing immediate feedback and adjusting the difficulty of tasks in real-time. Such training simulations are not only more engaging but also allow for the safe practice of complex tasks, thereby improving learning outcomes and operational preparedness without the associated risks of training in a physical environment.

As XR technologies continue to evolve, the integration of AI is set to revolutionize how warehousing operations are conducted. By enhancing image processing, improving human-robot interfaces, and creating dynamic training environments, AI is making XR applications more effective, efficient, and tailored to meet the growing demands of the logistics and supply chain sectors.

6.5.2 Trends in XR Hardware: The Shift from AR and VR to MR

As the boundaries between VR and AR continue to blur, the emergence of MR represents a significant shift in the landscape of immersive technologies, particularly in industrial applications and warehousing (Spadoni et al., 2023). This convergence is driven by the need to combine the immersive depth of VR with the contextual enhancement of AR, creating a more fluid and versatile user experience. High-resolution displays, sophisticated motion tracking, and more powerful processing capabilities allow MR devices to deliver rich, immersive experiences that were previously the sole domain of VR, while maintaining the real-world context provided by AR. Enhanced sensor technology and better integration of digital and physical data streams enable MR to provide interactive experiences that are both deep and responsive to the user's environment.

While AR has been transformative in many sectors, its limitations are becoming apparent, especially in industrial settings like warehousing. AR typically overlays digital information onto the real world but often struggles with depth perception and user disorientation during complex interactions. These limitations can hinder operational efficiency in environments where precision and context are critical. MR addresses these challenges by not only overlaying but also integrating digital objects into the physical world in a way that they are perceived as part of the environment, thus providing a more intuitive user interface. In warehousing, MR can revolutionize several core operations, from inventory management to training and maintenance. For instance, MR can enable warehouse workers to see digital overlays of inventory data directly on warehouse racks, guide them through optimal picking routes, and provide real-time information about potential hazards. The ability to switch between fully immersive (VR-like) and augmented (AR-like) experiences allows for flexible adaptations to different tasks, such as training in a fully controlled virtual space or completing maintenance with overlaid instructions on actual equipment.

The integration of VR and AR into a unified MR framework presents substantial opportunities but also significant challenges. Interoperability between different devices and software, user comfort, and the management of data privacy and security are critical hurdles that need addressing. Furthermore, as MR devices become more capable, ensuring they remain accessible and easy to use without extensive training remains a priority. As industries increasingly recognize the potential of MR to streamline operations and enhance worker productivity, the focus will likely shift toward creating more robust, user-friendly, and versatile MR solutions. In warehousing, this could mean the difference between a facility that struggles with efficiency and one that sets new standards for innovation and performance. The convergence of VR and AR into MR not only represents a technological evolution but also marks a shift in how industries will approach and solve real-world problems using digital solutions.

6.5.3 The Future of XR in Eye Implants

Advancements in meta surface technology are poised to revolutionize augmented reality applications by eliminating the need for bulky HMDs. According to Lan et al. (2019), metasurfaces enable the projection of digital information directly onto the retina through a contact lens display (CLD), which manages to display arbitrary patterns with high resolution and low noise without the need for additional optical elements like spatial light modulators (Lan et al., 2019). This technology facilitates a highly integrated and unobtrusive AR experience, offering a discreet and efficient alternative to traditional AR systems. These advancements could significantly enhance operational efficiency in environments like warehousing, allowing workers to access real-time data superimposed directly onto their natural vision, thus eliminating the physical encumbrance of traditional devices.

Looking toward the far future of XR, the potential integration of metasurface technology directly into the human body could lead to the development of eye implants capable of AR display. Such implants would render external devices entirely obsolete, providing a seamless interface between digital data and the physical world. The continuous miniaturization and integration of AR technology could lead to a scenario where digital enhancement of our sensory inputs is as natural and ubiquitous as wearing corrective lenses today. These future technologies would not only revolutionize how we interact with information-rich environments but also significantly expand the capabilities of human perception in everyday life.

6.5.4 Integrating WMS and IIoT in Warehousing: A Step Towards Industry 5.0

The integration of WMS with automation and the Industrial Internet of Things (IIoT) is an essential element in the ongoing evolution towards Industry 5.0, which emphasizes a human-centric approach where robots and automated systems collaborate more closely with human operators. This new paradigm seeks to enhance the synergy between human creativity and mechanical efficiency, elevating the roles that both play in the industrial ecosystem, particularly in warehousing operations. As described by Fatima et al. (2022), Industry 5.0 aims to maximize the benefits of human-machine interaction by maintaining a balance between ever-increasing powerful machinery and the productive abilities of human beings.

In the context of warehousing, the integration of WMS with IIoT facilitates a seamless flow of information across various platforms, enhancing the decision-making process and operational efficiency (Fatima et al., 2022). Robots and automated systems, equipped with AI and machine learning, can perform repetitive tasks with high precision, while humans manage more complex or sensitive decisions and oversight, leveraging the data-driven insights provided by IIoT. This collaboration is supported by sophisticated interfaces that allow humans and machines to interact intuitively, thereby minimizing errors and increasing productivity.

Looking further, the robust integration of these systems promises not only to streamline warehouse operations but also to adapt dynamically to changing demands and conditions. As IIoT devices become more interconnected, the data they collect and share will enhance the predictive capabilities of WMS, leading to better resource management, optimized workflows, and ultimately, a more responsive supply chain. This integration aligns with the goals of Industry 5.0 by not only optimizing the technical aspects of warehouse operations but also by enhancing the work environment for human operators, making it safer and more productive.

6.5.5 Future Prospects: The Integration of XR in Warehouse Management

The trajectory of XR in the realm of warehouse management is geared toward creating a more integrated and efficient operational environment through the strategic deployment of AI and MR. The advancements in AI-driven image processing for item identification and sorting have significantly minimized manual errors and streamlined inventory management. This fusion not only enhances the precision of operations but also optimizes the utilization of resources, thereby contributing to more effective warehouse logistics.

Transitioning to MR, this evolution from AR and VR stands as a critical innovation in immersive technology, tailored specifically for industrial applications. MR amalgamates the immersive essence of VR with the practical overlays of AR, crafting an interactive workspace where digital and physical realms converge seamlessly. This development overcomes previous barriers faced by AR technologies, such as depth perception and navigational disorientation, facilitating a more intuitive and productive user experience within the warehouse setting.

In conclusion, the ongoing integration of XR technologies is poised to significantly elevate the capabilities of warehouse operations. By harmonizing AI's analytical precision with MR's interactive enhancements, the future of warehouse management looks towards a more dynamic, efficient, and human-centric approach. This integration not only promises to refine operational processes but also sets a new standard for innovation in logistics and supply chain management, heralding a new era of industry-wide transformation.

6.6 Conclusion

The adoption and integration of XR technologies, including VR, AR, and MR, signify a transformative leap in the management of smart warehouses. These innovations are deeply rooted in the principles of Industry 4.0 and are evolving with the progressive ideals of Industry 5.0. They have redefined traditional warehousing operations, setting new benchmarks for efficiency, precision, and human-robot collaboration in the logistics sector.

VR has revolutionized training and layout planning within warehouses, offering immersive and interactive environments that enable risk-free exploration and operational refinement. AR has transformed critical logistics operations by overlaying digital information directly onto the physical world, thereby enhancing accuracy and efficiency at every step. MR, which blends the capabilities of AR and VR, allows warehouse operators to engage with both digital overlays and the real environment simultaneously, fostering informed decision-making and efficient task execution.

Despite the substantial benefits brought by XR technologies, their integration into warehousing faces several challenges. These include technological hurdles related to hardware design and system integration, financial constraints such as the high costs associated with advanced technologies, and human factors like workforce training and adaptation. Additionally, the imperative to secure sensitive operational data within XR platforms presents significant security challenges that require robust encryption and strict access controls to address.

Looking forward, the future of warehousing is poised for further transformation through the strategic integration of XR with AI and the IoT. This synergy is expected to enhance inventory management with superior predictive analytics and real-time data processing, revolutionizing how humans interact with automated systems. The promise of XR technologies to facilitate a collaborative, efficient, and adaptive warehouse environment is emerging, with AI and IoT amplifying the capabilities of XR to meet the dynamic demands of the global supply chain.

To fully realize the potential of XR in warehousing, it is crucial to focus on developing ergonomic and user-friendly interfaces that accommodate all users, regardless of their technical proficiency. This approach will help ensure that technological advancements enhance rather than replace human roles within warehouse operations. With ongoing innovation, strategic implementation, and effective training programs, XR technologies are set to optimize warehouse functionality, improve worker roles, and enhance the overall productivity and safety of the logistics industry. This journey highlights a future where digital enhancements and physical operations are seamlessly integrated, creating a more dynamic, efficient, and human-centric approach to warehouse management.

Declaration of Generative AI and AI-assisted Technology in the Writing Process

During the preparation of this manuscript, the author utilized ChatGPT for generating initial structure and refining specific text segments, and Napkin for minor image modifications. The author created and curated the core content, structure, and references, ensuring accuracy and alignment with the intended academic standards. All AI-generated content was thoroughly reviewed, edited, and validated by the author, who takes full responsibility for the final manuscript.

Acknowledgment

The author would like to sincerely thank Associate Professor Hao Yu at UiT The Arctic University of Norway for his invaluable guidance and mentorship during the preparation of this chapter. His extensive expertise in sustainable supply chain management, reverse logistics, and smart logistics in Industry 4.0/5.0 provided critical insights that shaped the exploration of XR technologies in warehousing. Dr. Yu's research and experience addressing real-world challenges through advanced analytics and optimization models were instrumental in enhancing the depth and quality of this work.

The author also acknowledges the Department of Civil and Industrial Engineering at the University of Pisa for providing the resources and environment necessary to complete this chapter. Finally, gratitude is extended to the internal and external reviewers for their constructive feedback, which greatly contributed to refining the manuscript.

References

Akbari, M., HA, N. and Kok, S. 2022. A systematic review of AR/VR in operations and supply chain management: maturity, current trends and future directions. *Journal of Global Operations and Strategic Sourcing*, 15, 534–565.

Akhmetov, T. and Varol, H. 2023. An Augmented Reality-Based Warning System for Enhanced Safety in Industrial Settings. *IEEE Transactions on Industrial Informatics*, 19, 7966–7977.

Blanco-Novoa, O., Fernandez-Carames, T.M., Fraga-Lamas, P. and Vilar-Montesinos, M.A. 2018. A practical evaluation of commercial industrial augmented reality systems in an industry 4.0 shipyard. *Ieee Access*, 6, 8201–8218.

De Guzman, J.A., Thilakarathna, K. and Seneviratne, A. 2019. Security and privacy approaches in mixed reality: A literature survey. *ACM Computing Surveys (CSUR)*, 52, 1–37.

Fatima, Z., Tanveer, M.H., Waseemullah, W., Zardari, S., Naz, L.F., Khadim, H., Ahmed, N. and Tahir, M. 2022. Production Plant and Warehouse Automation with IoT and Industry 5.0. *Applied Sciences*, 12 (4): 2053.

Glockner H., Jannek K., Mahn J., Theis B. Augmented Reality in Logistics: Changing the Way We See Logistics – A DHL Perspective. DHL Customer Solutions & Innovation, Troisdorf, Germany, 28 pp. (White paper, 2014).

Gonzalez-Franco, M., Pizarro, R., Cermeron, J., Li, K., Thorn, J., Hutabarat, W., Tiwari, A. and Bermell-Garcia, P. 2017. Immersive mixed reality for manufacturing training. *Frontiers in Robotics and AI*, 4, 3.

Grabowski, A. and Wodzyński, M. 2022. Utilisation of Games and Virtual Reality to Train Competences Relating to Industry 4.0. *Problems of Mechatronics. Armament, Aviation, Safety Engineering*, 13, 87–96.

Grosse, E., Sgarbossa, F., Berlin, C. and Neumann, W. 2023. Human-centric production and logistics system design and management: transitioning from Industry 4.0 to Industry 5.0. *International Journal of Production Research*, 61, 7749–7759.

Havard V., Trigunayat A., Richard K., Baudry D. Collaborative Virtual Reality Decision Tool for Planning Industrial Shop Floor Layouts. Procedia CIRP 81: 1295–1300.

Husár, J. and Knapčíková, L. 2021. Possibilities of using augmented reality in warehouse management: A study. *Acta logistica*, 8, 133–139.

Jumahat, S., Sidhu, M.S. and Shah, S.M. Pick-by-vision of augmented reality in warehouse picking process optimization–a review. 2022 IEEE International Conference on Artificial Intelligence in Engineering and Technology (IICAIET), 2022. IEEE, 1–6.

Kalinov, I., Trinitatova, D. and Tsetserukou, D. 2021. WareVR: Virtual Reality Interface for Supervision of Autonomous Robotic System Aimed at Warehouse Stocktaking. *2021 IEEE International Conference on Systems, Man, and Cybernetics (SMC)*, 2139-2145.

Kurschl, W., Pimminger, S., Schönböck, J., Augstein, M. and Altmann, J. 2021. Using mixed reality in intralogistics-are we ready yet? *Procedia Computer Science*, 180, 132–141.

Lan, S., Zhang, X., Taghinejad, M., Rodrigues, S., Lee, K.-T., Liu, Z. and Cai, W. 2019. Metasurfaces for Near-Eye Augmented Reality. *ACS Photonics*, 6, 864–870.

Limeira, M., Piardi, L., Cremer Kalempa, V., Schneider, A. and Leitão, P. Augmented reality system for multi-robot experimentation in warehouse logistics. Robot 2019: Fourth Iberian Robotics Conference: Advances in Robotics, Volume 1, 2020. Springer, 319–330.

Matsumoto, T., Kosaka, T., Sakurada, T., Nakajima, Y. and Tano, S. Picking Work using AR Instructions in Warehouses. 2019 IEEE 8th Global Conference on Consumer Electronics (GCCE), 15-18 Oct. 2019. 31–34.

Nagda, M.K., Sinha, S. and Poovammal, E. 2019. An augmented reality assisted order picking system using IoT. *International Journal of Recent Technology and Engineering*, 8, 744–749.

Papcun, P., Cabadaj, J., Kajati, E., Romero, D., Landryova, L., Vascak, J. and Zolotova, I. Augmented reality for humans-robots interaction in dynamic slotting "chaotic storage" smart warehouses. Advances in Production Management Systems. Production Management for the Factory of the Future: IFIP WG 5.7 International Conference, APMS 2019, Austin, TX, USA, September 1–5, 2019, Proceedings, Part I, 2019. Springer, 633–641.

Perotti S., Bastidas Santacruz R.F., Bremer P. and Beer J.E. Logistics 4.0 in warehousing: a conceptual framework of influencing factors, benefits and barriers. *Int. J. Logist. Manag.* 33 (5): 193–220.

Piardi, L., Kalempa, V.C., Limeira, M., De Oliveira, A.S. and Leitão, P. 2019. Arena—augmented reality to enhanced experimentation in smart warehouses. *Sensors*, 19, 4308.

Plakas, G., Ponis, S.T., Agalianos, K., Aretoulaki, E. and Gayialis, S.P. 2020. Augmented reality in manufacturing and logistics: Lessons learnt from a real-life industrial application. *Procedia Manufacturing*, 51, 1629–1635.

Singh, N. and Adhikari, D. 2023. AI and IoT: A future perspective on inventory management. *International Journal for Research in Applied Science and Engineering Technology*, 11, 2753–2757.

Spadoni, E., Bordegoni, M., Carulli, M. and Ferrise, F. 2023. Extended Reality In Industry: Past, Present And Future Perspectives. *Proceedings of the Design Society*, 3, 1845–1854.

Speicher, M., Hall, B.D. and Nebeling, M. What is mixed reality? Proceedings of the 2019 CHI conference on human factors in computing systems, 2019. 1–15.

Tang, Y.M., Kuo, W.T. and Lee, C. 2023. Real-time Mixed Reality (MR) and Artificial Intelligence (AI) object recognition integration for digital twin in Industry 4.0. *Internet of Things*, 23, 100753.

Yigitbas E., Jovanovikj I., Scholand J., Engels G. VR Training for Warehouse Management. In: Proc. VRST '20 (Virtual Event, Canada). Article 78, pp. 1–3.

7 Circular Economy Principles and Sustainable Logistics Practices in Industry 5.0

Al-Amin Abba Dabo[1*], Bright Amajuoyi[2] and F. Frank Alparslan[3]

7.1 Introduction to Circular Economy and Industry 5.0

The Circular Economy (CE) represents a transformative approach to sustainable business practices and economic development, fundamentally shifting away from the traditional linear model characterised by the "take, make, dispose" paradigm. Instead, CE emphasises a regenerative and restorative system where resources are continually reused, thereby tackling global challenges such as resource scarcity, environmental degradation, and unsustainable growth by redefining how resources are valued and utilised in the economy (Chen et al., 2020). The focus is on designing waste out of the system, regenerating natural ecosystems, and maximising resource efficiency by keeping materials in circulation at their highest value for as long as possible (Bressanelli et al., 2022). CE challenges businesses to redesign their products and processes to ensure that value is retained within the system, whether through recycling, reuse, or the use of renewable energy sources. This shift not only reduces the environmental impact of production but also creates new opportunities for business models that align with sustainability objectives and societal expectations (Dabo & Hosseinian-Far, 2023).

Moreover, CE promotes a multi-dimensional approach that integrates environmental, economic, and social benefits. Environmentally, it significantly reduces the volume of waste generated and prevents pollution, while economically, it contributes to cost savings through resource efficiency and the creation of new markets for recycled and refurbished products (Fera et al., 2023). Socially, CE fosters a more equitable distribution of resources by prioritising regenerative practices that benefit communities and minimise harm to natural ecosystems. In contrast to the short-term thinking that underpins the linear model, CE introduces a long-term perspective that values the interconnectedness of natural, economic, and social systems, creating a more resilient and adaptive industrial ecosystem. By emphasising principles such as resource efficiency, renewable resource use, and product life extension, CE aims to create a closed-loop system where waste

[1-3] Department of Business, Systems and Operations, University of Northampton, Northampton, United Kingdom.
* Corresponding author: Al-Amin.Dabo@northampton.ac.uk

DOI: 10.1201/9781003466451-7

is minimised, and materials are continually cycled back into production (Calon, 2020; Bressanelli et al., 2022).

Building on the advancements of Industry 4.0, Industry 5.0 introduces a significant shift by reintegrating the human element into industrial processes. While Industry 4.0 primarily focused on automation, real-time data collection, and digital technology integration, Industry 5.0 emphasises collaboration between human resourcefulness and machine precision, thereby fostering an environment that supports innovation, creativity, and personalised production. This human-centric approach adds a crucial dimension to smart manufacturing, emphasising worker welfare, societal needs, and a balanced relationship between technology and humans (Narula et al., 2023). It aims to create a synergistic relationship that enhances productivity while simultaneously catering to the unique needs and abilities of the human workforce, ensuring that technological advancements do not undermine, but rather enhance, the quality of human work. This transition to Industry 5.0 represents a major opportunity for the logistics and supply chain sector, promoting resilience, sustainability, and social responsibility alongside operational efficiency (Zhang et al., 2023).

The logistics sector, which has traditionally been associated with high resource consumption and environmental degradation, is particularly poised to benefit from the integration of CE principles with Industry 5.0 technologies. The application of advanced technologies, such as AI, IoT, blockchain, and robotics, in logistics not only enhances operational efficiency but also supports the adoption of circular practices, such as product life extension and resource optimisation (Dabo & Hosseinian-Far, 2023). AI can optimize delivery routes, improve warehouse management, and enhance decision-making processes, which leads to reduced fuel consumption, decreased waste, and overall improved resource use (Latha & Patil, 2024). Similarly, IoT sensors provide real-time data on the condition of goods, enabling precise monitoring and timely intervention to prevent losses and ensure quality control. These technologies contribute to achieving the key objectives of CE by ensuring that materials are used efficiently, and waste is minimised throughout the logistics process (Chauhan et al.,2022; Bag et al., 2020).

However, the transition from Industry 4.0 to Industry 5.0 also highlights the importance of reintroducing the human factor into the equation. Unlike the purely technology-driven focus of Industry 4.0, which prioritised efficiency and productivity through automation, Industry 5.0 ensures that technological advancements serve broader societal values, such as human welfare, inclusivity, and environmental sustainability (Toth et al., 2023). The emphasis on human-machine collaboration transforms the logistics sector into one that not only strives for operational excellence but also prioritises worker safety, satisfaction, and engagement (Narula et al., 2023). This human-centered approach is essential for addressing the complexities of modern supply chains, where diverse stakeholders and variables require adaptive solutions that balance efficiency with social responsibility.

The integration of CE principles with Industry 5.0 technologies thus creates a powerful framework for fostering sustainable logistics practices. Circular economy strategies in logistics, such as reverse logistics and closed-loop supply chains, aim to reduce resource consumption, minimise waste, and maximise resource efficiency (Mishra et al., 2023). By embedding these circular practices within logistics, companies can enhance the sustainability of their supply chains while also generating economic value through resource recovery (Dabo & Hosseinian-Far, 2023). The role of Industry 5.0 in enhancing these strategies is pivotal, as it incorporates a human-centric approach that leverages both technological precision and human creativity, ensuring that logistics operations are not only efficient but also sustainable and adaptive to changing societal needs (Fera et al., 2023). This synergy creates a comprehensive framework for sustainable logistics practices that enhance resource efficiency, reduce waste, and promote resilience (Ghoreishi & Happonen, 2020; Sarkar et al., 2024).

7.2 Integration of Circular Economy Principles in Logistics

The integration of Circular Economy (CE) principles in logistics has increasingly become a focal point for improving the efficiency and sustainability of supply chain operations. Logistics, as a fundamental component of the supply chain, plays an important role in the movement, storage, and management of resources, making it particularly well-suited to the adoption of CE principles that emphasise minimising waste and maximising resource use. The traditional linear logistics model, which prioritises speed and cost, often overlooks the environmental consequences associated with excessive resource consumption, emissions, and waste. By shifting to a circular approach, logistics can significantly reduce its environmental footprint while also unlocking new value through more efficient and sustainable resource management (Owen, 2024).

In the context of logistics, CE principles manifest through practices such as reverse logistics, closed-loop supply chains, and remanufacturing. Reverse logistics is a critical aspect of the CE, involving the process of returning products from consumers back to manufacturers for recycling, refurbishing, or remanufacturing. It not only enables companies to recover value from returned goods but also reduces the volume of waste sent to landfills, contributing to environmental sustainability (Borumand & Nookabadi, 2024). For instance, companies like Apple have successfully implemented reverse logistics systems to recover valuable components from old electronic devices, thus reducing their dependency on raw materials and minimising waste (Szmelter-Jarosz et al., 2024). Similarly, closed-loop supply chains seek to create a continuous cycle of product use by ensuring that end-of-life products are reintegrated into the supply chain as raw materials or refurbished products, thus creating a loop that minimises waste and maximises the value extracted from each product life cycle (Owen, 2024).

One of the significant challenges in logistics is managing the balance between cost, efficiency, and environmental impact. Traditional logistics networks are designed primarily for forward flows of goods, focusing on delivering products from manufacturers to customers as quickly and cost-effectively as possible. However, incorporating CE practices such as reverse logistics into these networks requires rethinking the entire logistics model, including reconfiguring transportation routes, redesigning warehouse layouts, and adopting new performance metrics that account for environmental and social impacts (ElMenshawy et al., 2023). Although such changes may initially increase operational complexity and costs, the long-term benefits of enhanced sustainability, resource recovery, and customer loyalty can outweigh these initial challenges (Vasiliauskas et al., 2024).

Moreover, collaboration across the supply chain is vital for the successful integration of CE principles into logistics. Manufacturers, logistics service providers, retailers, and consumers all have roles to play in ensuring that products are designed, used, and disposed of in ways that support circularity. For example, manufacturers can design products with modular components that facilitate easy disassembly and recycling, while logistics service providers can invest in technology and infrastructure to support efficient reverse logistics processes. Consumers, on the other hand, must be educated and incentivised to return products at the end of their life cycle, thereby enabling the continuous flow of resources within the supply chain (Shin et al., 2024).

Technology plays an instrumental role in facilitating the integration of CE principles in logistics. Innovations such as the Internet of Things (IoT), Artificial Intelligence (AI), and blockchain have the potential to significantly enhance the efficiency and effectiveness of circular logistics operations. IoT devices, for instance, enable real-time tracking of goods, providing valuable data that can be used to optimize transportation routes, monitor the condition of goods, and ensure timely and efficient reverse logistics processes. This real-time visibility helps reduce waste by ensuring that products are not lost, damaged, or delayed during transit, thereby maintaining their value for reuse or recycling (ElMenshawy et al., 2023).

AI, on the other hand, can be employed to optimise decision-making processes in logistics, such as determining the most efficient routes for transporting returned goods, predicting demand for refurbished products, and optimising inventory management to reduce overproduction and waste. By leveraging AI for predictive analytics, companies can anticipate the flow of returned goods and plan their logistics operations, accordingly, ensuring that resources are utilised efficiently and waste is minimised (Vasiliauskas et al., 2024). Blockchain technology, with its capability to provide transparent and immutable records of transactions, can also play a key role in enhancing trust and traceability in circular supply chains. By providing a transparent record of a product's journey through the supply chain, blockchain can help ensure that returned goods are handled appropriately, recycled materials are not mixed with virgin materials, and consumers have access to information about the sustainability of the products they purchase (Shin et al., 2024).

7.3 Industry 5.0: Enhancing Circularity in Logistics

The emergence of Industry 5.0 presents a significant evolution in the industrial landscape, emphasising human-centricity, resilience, and sustainability (Nahavandi, 2019; Martini et al., 2024). Unlike its predecessor, Industry 4.0, which focused primarily on the automation and digitalisation of processes through advanced technologies like the Internet of Things (IoT), Artificial Intelligence (AI), and big data, Industry 5.0 shifts the focus towards collaboration between humans and machines. This paradigm shift is particularly relevant to the logistics sector, where the integration of Circular Economy (CE) principles can be significantly enhanced by harnessing the power of human creativity, cognitive abilities, and technological precision (Narula et al., 2023; Mahiri et al., 2024). By positioning humans at the center of innovation and value creation, Industry 5.0 provides new opportunities for addressing the environmental, social, and economic challenges faced by the logistics industry in its quest to adopt CE practices (Nahavandi, 2019).

A key characteristic of Industry 5.0 is the emphasis on personalisation and customisation, which can contribute to the CE in logistics by optimising resource use and reducing waste. For instance, mass customisation can be enabled through collaborative robotics (cobots), which work alongside human workers to manufacture products that meet specific consumer requirements (Raffik et al., 2023). In the logistics context, this means that operations can be tailored to optimise the flow of goods and reduce inefficiencies. Customisation of packaging, routing, and inventory management, for example, allows logistics service providers to minimise the use of resources, reduce the volume of waste generated, and ensure that products are delivered in a way that maximises their value and lifespan (Mekkunnel, 2019; Rahman et al., 2024).

Incorporating the principles of Industry 5.0 into logistics also allows for the enhancement of remanufacturing and refurbishing processes, key components of the CE. The integration of human expertise with the precision of robotics and AI can lead to significant improvements in the efficiency and quality of these processes. For example, the disassembly of returned products is a labour-intensive task that requires careful handling to ensure that components are not damaged and can be reused. In Industry 5.0, human workers, supported by cobots, can perform this task more effectively, leveraging the strengths of both human judgment and robotic precision. This results in higher rates of component recovery, reduced waste, and increased value retention, thereby contributing to the sustainability of the logistics sector (Raffik et al., 2023; Mahiri et al., 2024).

Industry 5.0 also supports the integration of CE principles in logistics through the use of AI-driven decision support systems, which enhance human decision-making capabilities in areas such as route optimisation, inventory management, and demand forecasting. AI systems can analyse vast amounts of data to provide logistics managers with insights into the most efficient routes for transporting goods, the optimal levels of inventory to maintain, and the best times to initiate reverse logistics processes. These insights help reduce waste, minimise energy consumption, and ensure that resources are used as efficiently as possible. At the same time, human expertise is critical in interpreting these insights and making decisions that take into account factors such as customer preferences, regulatory requirements, and ethical considerations, which may not be fully captured by AI algorithms (Bhowmik et al., 2023; Martini et al., 2024).

Another important aspect of Industry 5.0 is its focus on resilience, which aligns with the CE's goal of creating a sustainable and adaptive supply chain. The logistics sector is particularly vulnerable to disruptions, such as those caused by natural disasters, geopolitical tensions, or pandemics. By incorporating CE principles, such as resource recovery and closed-loop supply chains, logistics companies can enhance their resilience, reducing their dependence on finite resources and minimising the impact of disruptions on their operations. Industry 5.0 supports this resilience by fostering a collaborative environment in which humans and machines work together to develop innovative solutions to complex problems, such as designing more adaptive logistics networks that can quickly respond to changing conditions and ensure the continuous flow of goods, even in the face of disruptions (Nguyen et al., 2023; Mahiri et al., 2024).

Human-centricity, another cornerstone of Industry 5.0, also has significant implications for the integration of CE principles in logistics. A human-centric approach recognises the importance of human well-being and aims to create a work environment that is safe, fulfilling, and conducive to innovation (Ivanov, 2023). In the logistics sector, this means prioritising the health and safety of workers, reducing the physical demands of tasks through automation and robotics, and providing opportunities for upskilling and professional development. By creating a more supportive and empowering work environment, logistics companies can foster a culture of sustainability in which employees are more engaged and motivated to contribute to CE initiatives. For example, workers who are empowered to identify and address inefficiencies in logistics processes are more likely to develop innovative solutions for reducing waste, optimising resource use, and enhancing the sustainability of operations (Grosse et al., 2023; Yaqot, 2024).

Moreover, the integration of Industry 5.0 principles into logistics can facilitate greater collaboration across the supply chain, which is essential for the successful adoption of CE practices. Collaboration between manufacturers, logistics service providers, retailers, and consumers are necessary to ensure that products are designed, used, and disposed of in a way that supports circularity (Saberi et al., 2019). Industry 5.0 technologies, such as AI and blockchain, can enhance collaboration by providing real-time visibility into the flow of goods, enabling more effective communication and coordination between supply chain partners. For example, blockchain can be used to create a transparent and immutable record of a product's journey through the supply chain, ensuring that all parties have access to the same information and can work together to achieve common sustainability goals. This level of transparency is particularly important for building trust between supply chain partners and ensuring that CE practices, such as the use of recycled materials or the return of end-of-life products, are implemented effectively (Saberi et al., 2019; Pizon et al., 2023).

In conclusion, Industry 5.0 represents a significant opportunity for enhancing the integration of Circular Economy principles in logistics. By emphasising human-centricity, resilience, and collaboration, Industry 5.0 enables the logistics sector to move beyond the limitations of automation and digitalisation, harnessing the combined power of human creativity and technological precision to create more sustainable and adaptive supply chain operations. The collaborative relationship between humans and machines in Industry 5.0 allows for the optimisation of logistics processes, the enhancement of remanufacturing and refurbishing activities, and the development of innovative solutions to complex challenges. By fostering a culture of sustainability and leveraging advanced technologies to enhance collaboration across the supply chain, Industry 5.0 provides a powerful framework for transforming the logistics sector into a key driver of the Circular Economy. The following section will consider the practical implementation of these concepts in logistics operations and discuss case studies that highlight the successful integration of CE and Industry 5.0 principles in the real world.

7.4 Case Studies: Integrating Circular Economy and Industry 5.0 in Logistics

To understand how the principles of Industry 5.0 and Circular Economy (CE) can be effectively integrated into logistics, it is beneficial to examine real-world examples that illustrate the practical implementation of these concepts. The following case studies highlight how companies have adopted innovative practices that align with the goals of Industry 5.0—focusing on human-centric approaches, resilience, and sustainability—while advancing the principles of CE within their logistics operations.

7.4.1 DHL's Sustainable Packaging Solutions

DHL, a global leader in logistics and supply chain management, has implemented several initiatives to promote sustainability and circularity across its operations. One notable example is its sustainable packaging solutions program, which aligns with both Circular Economy (CE) and Industry 5.0 principles by integrating advanced technologies and human ingenuity to minimise waste and improve resource efficiency. As part of this initiative, DHL introduced AI-driven packaging optimization tools that analyse data on product dimensions, material use, and shipping routes. This analysis enables the reduction of excess packaging material, ultimately leading to significant savings and less environmental impact. This initiative not only supports CE principles by minimising resource consumption and waste generation but also incorporates the human element by involving employees in identifying opportunities for packaging improvements based on their operational experience (DHL, 2023a; DHL, 2023b).

7.4.1.1 Integration of Industry 5.0 Technologies

Through the application of Industry 5.0 technologies such as machine learning and collaborative robotics, DHL has enhanced the packaging customisation process, reducing the environmental impact of logistics. Collaborative robots, or cobots, work alongside DHL employees to handle complex packaging tasks, allowing for more precise packaging customisation that reduces material waste. This collaboration allows employees, equipped with insights provided by AI, to play a crucial role in fine-tuning packaging strategies to meet specific customer requirements. By combining machine intelligence with human creativity, DHL ensures optimal material use while simultaneously improving worker safety and well-being, as cobots handle physically demanding tasks and reduce the physical strain on human workers (DHL, 2022; DHL, 2023a; DHL, 2023b). Figure 7.1 represents the packaging optimization workflow.

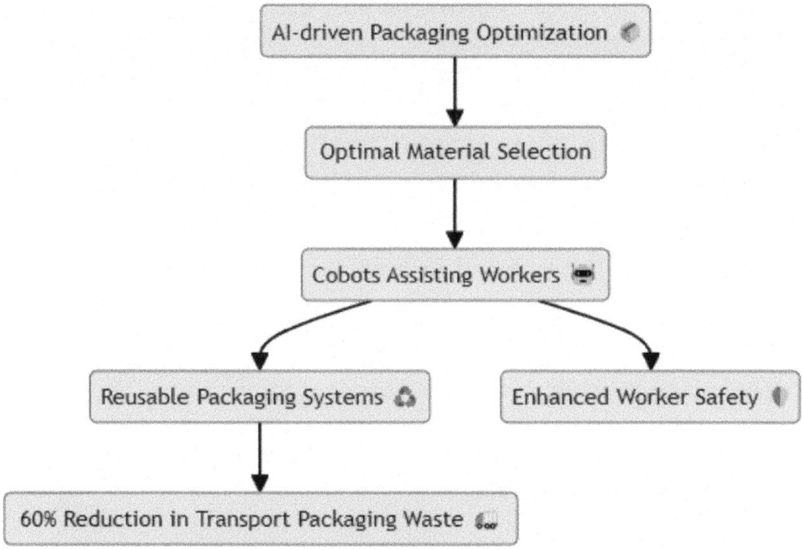

Fig. 7.1 DHL packaging optimisation workflow (*Source:* Authors).

A significant outcome of this initiative is the reduction in packaging waste by 25%, achieved through data-driven analysis coupled with employee input on improvements. This demonstrates how integrating human insight and advanced technologies can create a more efficient and environmentally friendly supply chain. Moreover, DHL's reusable packaging systems have led to a 60% reduction in transport packaging waste, providing clear evidence of the tangible benefits of integrating CE principles into logistics. This dual approach—focusing on both improving single-use packaging and shifting toward circular, reusable alternatives—positions DHL at the forefront of sustainable logistics practices (DHL, 2022; DHL, 2023a; DHL, 2023b).

7.4.1.2 Optimised, Sustainable Packaging and Advanced Consumables

DHL's sustainable packaging solutions focus on reducing waste by optimising the use of consumables and incorporating recycled and recyclable materials. The use of Advanced Sustainable Warehouse Consumables involves initiatives like repurposing used inbound boxes to become void fill, adopting box-on-demand technologies, and replacing traditional pallet wraps with reusable wraps that can endure over 250 cycles. These innovations help DHL achieve significant reductions in material waste, while also ensuring operational efficiency. Technologies such as linerless labels and re-usable pallet wraps further reduce the consumption of single-use materials, decreasing DHL's overall carbon footprint (DHL, 2022).

7.4.1.3 Circular Packaging and Sustainable Secondary Packaging

The emphasis on Circular Packaging Solutions is a critical aspect of DHL's strategy to reduce transport packaging waste. The company uses returnable packaging and embraces the concept of "naked products," which means reducing the use of packaging materials to an absolute minimum where possible. Additionally, Sustainable Secondary Packaging is designed to replace fossil-based plastics with alternatives like post-consumer recycled (PCR) and bio-based materials, thereby reducing the reliance on virgin materials and helping clients meet their sustainability targets. DHL also actively consults, sources, and executes sustainable packaging for clients, including those in the e-commerce sector, contributing to a broader impact in sustainable supply chain management (DHL, 2023a; DHL, 2023b).

7.4.1.4 Green Single-Use Packaging and Reusable Solutions

DHL employs a strategic dual approach to green packaging, simultaneously improving existing single-use packaging and implementing reusable solutions. The improved single-use packaging materials—such as thinner pallet wraps with high recycled content—are designed to be more sustainable while still fulfilling the necessary performance requirements for logistics operations. Meanwhile, reusable packaging solutions such as plastic boxes and bags have an impressive ecological breakeven point after a few uses and can be reused for more than 50 cycles. These solutions help reduce the use of virgin plastic, cut CO_2 emissions, and contribute to significant cost savings. For example, the use of reusable pallet wraps not only cuts plastic use but also achieves a financial break-even after approximately 65 cycles, making them both environmentally and economically viable (DHL, 2023a).

7.4.1.5 Service Logistics Smart Circle

In addition to packaging optimisation, DHL's Service Logistics Smart Circle integrates circularity into its logistics services through a comprehensive approach that extends product lifecycles. This system allows for the return, inspection, recovery, and reuse of products. Customers initiate returns via a self-service portal, and DHL's technical services team inspects and repairs or remanufactures returned items as needed. This allows products to re-enter the supply chain either for direct reuse or for resale, thus promoting a circular model that not only reduces waste but also maximises the value extracted from each product lifecycle (DHL, 2023a; DHL, 2023b).

7.4.1.6 Practical Implications of DHL's Sustainable Packaging Solutions

The practical implications of DHL's sustainable packaging solutions extend across environmental sustainability, operational efficiency, and economic savings. Environmentally, DHL's initiatives significantly cut down on material usage, waste generation, and carbon emissions, aligning with global sustainability goals. Operationally, the use of advanced technologies such as AI-driven box-on-demand and cobots in packaging tasks optimises the logistics process, ensuring that only the necessary materials are used and that packaging is done efficiently, thus reducing excess material and improving transport efficiency. Economically, adopting reusable packaging systems brings long-term cost savings by reducing the frequency of purchasing new materials and lowering waste disposal costs. Furthermore, DHL's use of AI to analyse product dimensions and determine optimal packaging helps maximise space utilisation during transportation, leading to reduced transportation costs and decreased fuel consumption (DHL, 2023a; DHL, 2023b).

The integration of AI-driven packaging optimization leads to a 10% reduction in material consumption, contributing directly to both cost savings and transportation efficiency gains. Collaborative robots, by reducing manual handling, enhance worker safety by 20% and boost worker productivity by 15%, showcasing the clear benefits of human-machine collaboration. Moreover, the reusable packaging systems result in a 60% reduction in packaging waste and a 30% reduction in costs related to packaging materials, reinforcing the value of circular economy practices.

In conclusion, DHL's sustainable packaging process represents a comprehensive effort to align logistics operations with CE and Industry 5.0 principles. Through AI, cobots, and a strong emphasis on reusability, DHL minimises packaging waste, optimises resource use, and reduces the environmental footprint of its supply chain operations. These initiatives not only improve operational efficiency and sustainability outcomes but also provide significant economic benefits, underscoring DHL's leadership in pioneering sustainable logistics solutions. The integration of advanced technologies and human insight within DHL's packaging processes sets an industry standard for how logistics providers can adopt sustainable practices without sacrificing efficiency or performance.

7.4.2 IKEA's Circular Supply Chain and Reverse Logistics

IKEA, the multinational furniture retailer, has committed to becoming a fully circular business by 2030, and its logistics operations play a critical role in achieving this goal. To advance this commitment, IKEA has adopted Circular Economy (CE) principles throughout its supply chain, including designing products for disassembly, reusing materials, and implementing reverse logistics to effectively handle returns and end-of-life products. The company's approach exemplifies the integration of Industry 5.0 concepts, such as human-centric design, AI-driven decision-making, and advanced automation, to enhance logistics processes for greater sustainability and resilience (IKEA, 2022). This initiative not only aligns with environmental goals but also aims to enhance customer satisfaction through a seamless and sustainable product lifecycle.

7.4.2.1 Reverse Logistics and Human-Machine Collaboration

To facilitate effective reverse logistics, IKEA has implemented AI-enabled systems that track and manage the flow of returned products and materials in real-time. These AI systems provide insights into inventory levels, product conditions, and other critical data points, allowing logistics managers to determine whether items should be refurbished, recycled, or resold. This AI-human collaboration exemplifies how technology can empower human decision-making by providing accurate data to inform key choices, thereby maximising material recovery and supporting sustainable supply chain practices. Workers assess the condition of returned products and bring their expertise into deciding on the appropriate course of action based on the product's quality and customer requirements, highlighting the integration of human judgment and AI (Malmgren and Larsson, 2020; Gomes, 2020).

The use of collaborative robots (cobots) in the disassembly process further emphasises IKEA's commitment to Industry 5.0's human-centric approach. Cobots assist workers by handling repetitive and physically demanding tasks, such as unscrewing and sorting components. By taking over these more strenuous duties, cobots reduce physical strain on employees and create a safer work environment. This combination of human expertise with technological precision enables IKEA to create a circular logistics system that is both resilient and sustainable. It also reflects IKEA's broader goal to make logistics operations more adaptive and efficient while prioritising the well-being of its workforce. Figure 7.2 captures the flow of returned products through IKEA's reverse logistics process, illustrating the role of AI in assessment, cobot involvement in disassembly, and how refurbished products are subsequently resold or recycled (Malmgren and Larsson, 2020; Gomes, 2020).

7.4.2.2 Data-Driven Insights and Reverse Logistics Improvements

IKEA has also taken significant steps to optimize its reverse logistics operations by developing and implementing data-driven insights to monitor return trends, predict reverse flows, and optimize its logistics network. This data-centric approach began with the development of standardised templates and tools for

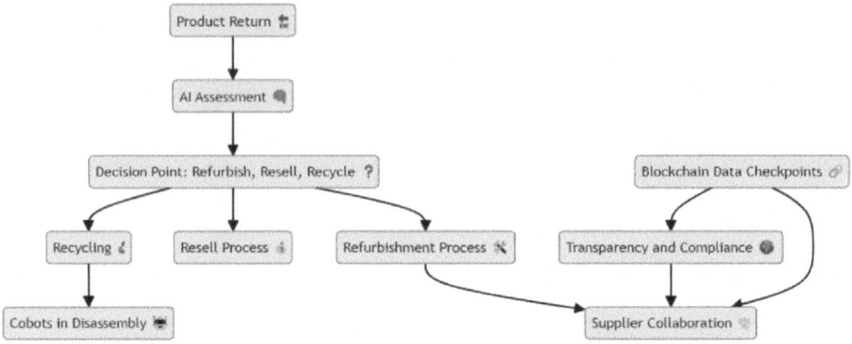

Fig. 7.2 IKEA closed-loop supply chain framework (*Source:* Authors).

systematically aggregating weekly return data from stores such as Alfragide in Portugal. By analysing return behaviours over time, IKEA identified which product categories, or Home Furnishing Businesses (HFBs), were most prone to returns and why. Categories such as bedroom furniture, kitchens, sofas, and mattresses collectively accounted for over 50% of total returns, highlighting the importance of understanding customer buying behaviours and the conditions leading to returns (Malmgren and Larsson, 2020; Gomes, 2020).

The detailed analysis of return data also led to the identification of specific causes for returns, including customer dissatisfaction due to a mismatch between expectations and the delivered product, changes of mind, or sales process errors. By using these insights, IKEA implemented improvements in product descriptions, packaging, and customer communication to mitigate future returns. Moreover, the ability to predict return patterns using aggregated data allows IKEA to optimize inventory management, ensuring that refurbished items re-enter the supply chain efficiently. This proactive approach helps IKEA maintain high levels of customer satisfaction while minimising costs associated with reverse logistics (Malmgren and Larsson, 2020; Gomes, 2020).

7.4.2.3 Collaboration with Supply Chain Partners

A critical element of IKEA's strategy to establish a circular supply chain is its extensive collaboration with suppliers and logistics service providers. This collaboration is made possible and enhanced by blockchain technology, which provides a transparent and immutable record of materials and products as they move through the supply chain. Blockchain technology ensures adherence to CE practices throughout the product lifecycle by enabling a verifiable chain of custody for materials, which is crucial in creating trust among supply chain partners and validating sustainable practices. By using blockchain to track the flow of materials from end-of-life products back into production, IKEA ensures that valuable resources are reintegrated into the supply chain, minimising waste and reducing the need for new raw materials (Malmgren and Larsson, 2020; Gomes, 2020).

This collaboration also extends to working closely with suppliers to ensure that products are designed for disassembly and ease of material recovery. For example, IKEA works with its suppliers to standardise components and use materials that can be easily recycled or reused. This approach not only reduces the complexity of handling returned items but also increases the efficiency of the reverse logistics process, allowing materials to be processed and reintegrated into the production cycle more effectively (Malmgren and Larsson, 2020; Gomes, 2020).

7.4.2.4 Practical Implications for IKEA's Sustainability and Circular Supply Chain

The practical implications of IKEA's circular supply chain and reverse logistics initiatives are multifaceted, involving environmental sustainability, operational efficiency, and enhanced customer satisfaction. By designing products specifically for disassembly and reusability, IKEA has embraced the core principles of the Circular Economy, reducing the waste generated from end-of-life products and ensuring that valuable materials are consistently recovered and reused. The integration of AI-enabled decision-making in managing product returns allows IKEA to optimize the value of each returned product, whether through refurbishment, resale, or recycling. This approach not only reduces landfill waste but also cuts costs associated with material procurement and disposal (Malmgren and Larsson, 2020; Gomes, 2020).

Moreover, the use of collaborative robots in the disassembly process highlights IKEA's commitment to a human-centric Industry 5.0 approach. Cobots relieve employees from labour-intensive tasks, which improves working conditions and reduces injury risks, while allowing employees to focus on decision-making and higher-value activities such as quality assessment and customer service. This balanced integration of human and machine roles improves overall productivity and helps IKEA manage the large volume of returns generated by its liberal 365-day return policy more effectively (Malmgren and Larsson, 2020; Gomes, 2020).

Another significant implication of IKEA's approach is the enhancement of supplier relationships through blockchain-facilitated transparency. By providing a reliable and auditable record of material flows, blockchain technology supports stronger supplier collaboration, ensuring that all partners adhere to sustainable practices throughout the product life cycle. This transparency helps IKEA not only meet its sustainability targets but also assures customers that the company is maintaining high standards in its commitment to environmental responsibility (Malmgren and Larsson, 2020; Gomes, 2020).

In conclusion, IKEA's integration of advanced technologies and circular principles in its supply chain showcases the company's dedication to becoming a circular business by 2030. Through the use of AI-driven decision-making, cobots for labour-intensive tasks, and blockchain to foster transparency and collaboration,

IKEA has developed a reverse logistics system that is not only efficient and cost-effective but also aligned with sustainable practices. These initiatives demonstrate how Industry 5.0 concepts can be applied to logistics to create systems that are not only technologically advanced but also centered on enhancing human capabilities and environmental stewardship. This holistic approach positions IKEA as a leader in integrating sustainability into its logistics operations, setting a benchmark for how companies can innovate to meet the challenges of a circular economy.

7.4.3 Maersk's Human-Centric Automation in Logistics

Maersk, a leading global container shipping company, has embraced the principles of Industry 5.0 to enhance its logistics operations and support circular economy (CE) goals. This approach focuses on human-centric automation, ensuring that employee well-being is prioritised alongside efficiency improvements. By introducing autonomous vehicles and AI-powered systems in its port terminals, Maersk has enhanced container handling, reduced fuel consumption, and minimised emissions. These technologies not only streamline logistics processes but also align with CE principles, minimising resource use and reducing waste generation, thus contributing to a more sustainable supply chain (Maersk, 2022).

7.4.3.1 Autonomous Vehicles and Employee Empowerment

The deployment of autonomous vehicles in Maersk's port terminals has been a significant step toward achieving more efficient logistics. These vehicles have successfully reduced fuel consumption by 15%, a direct benefit in terms of emission reductions and resource conservation. Furthermore, AI-powered systems have been implemented to manage inventory levels effectively, resulting in a 10% reduction in overstock and associated waste (Dagar et al., 2024). These efficiencies support Maersk's broader goals of resource optimisation and CE. However, Maersk's human-centric automation approach goes beyond just deploying advanced technology; it emphasises the collaboration between humans and machines. Employees at port terminals are trained to operate and interact with autonomous vehicles, allowing them to fully leverage the advantages of automation while retaining control over critical aspects of logistics. This balance ensures that workers are empowered to make strategic decisions that impact the operation, while the automation takes care of repetitive and mundane tasks. This integration has created a resilient logistics environment, one where technology supports the workforce rather than replacing it.

The benefits of this human-machine collaboration are manifold. By alleviating the physical strain and reducing the risks associated with heavy container movements, Maersk has improved workplace safety and enhanced the overall job satisfaction of its employees. In turn, this human-centric approach has contributed to the sustainability of Maersk's operations by promoting responsible resource management and reducing environmental impacts such as emissions (Dagar et al., 2024).

7.4.3.2 Human-Centric Automation and Circular Economy Integration

Industry 5.0's emphasis on a human-centric design is evident in Maersk's approach to logistics. Autonomous vehicles and AI are employed not only for operational gains but also for advancing CE goals. By minimising overstock and managing precise inventory control, AI-driven logistics ensures that resources are utilised efficiently, and waste is minimised, directly contributing to the CE (Maersk, 2022).

The training programs provided to terminal workers to operate autonomous vehicles are a vital part of Maersk's human-centric strategy. These initiatives ensure that employees are not left behind in the technological transition but are instead provided with opportunities to upgrade their skills and advance their careers. By involving workers in the technology adoption process, Maersk is enhancing their resilience to change and creating a more adaptive workforce. Employees are empowered to engage with technology, and their experience and judgment are integrated into the decision-making process, leading to more effective and sustainable logistics operations (Dagar et al., 2024).

Figure 7.3 illustrates Maersk's implementation of human-centric automation within its port logistics operations. It highlights the role of autonomous vehicles in significantly reducing fuel consumption and how AI-powered inventory systems effectively support workers in focusing on higher-level strategic decision-making. Additionally, it emphasises the synergistic collaboration between human operators and advanced automated technologies, showcasing how this integration not only optimises efficiency but also enhances the resilience and sustainability of the entire logistics process.

7.4.3.3 AI and Machine Learning for Continuous Improvement

Maersk's integration of AI and machine learning (ML) is a core part of its strategy for continuous improvement in logistics, aligned with Industry 5.0 and CE objectives. AI systems are implemented to offer real-time insights into container movement, fuel consumption, and other critical metrics, which allows logistics managers to make more informed decisions. For example, AI-driven predictive analytics can anticipate potential disruptions in supply chains, allowing Maersk to adapt in advance, avoiding delays, and optimising fuel usage and emissions (Maersk, 2022; Dagar et al., 2024).

The AI applications also include dynamic inventory management, which minimises waste and enhances resource efficiency by predicting inventory needs with greater accuracy. By accurately forecasting the demand and ensuring optimal stock levels, Maersk significantly reduces the risks associated with overstock and understock, thereby aligning operations closely with CE principles of reducing waste and improving the utility of resources (Dagar et al., 2024).

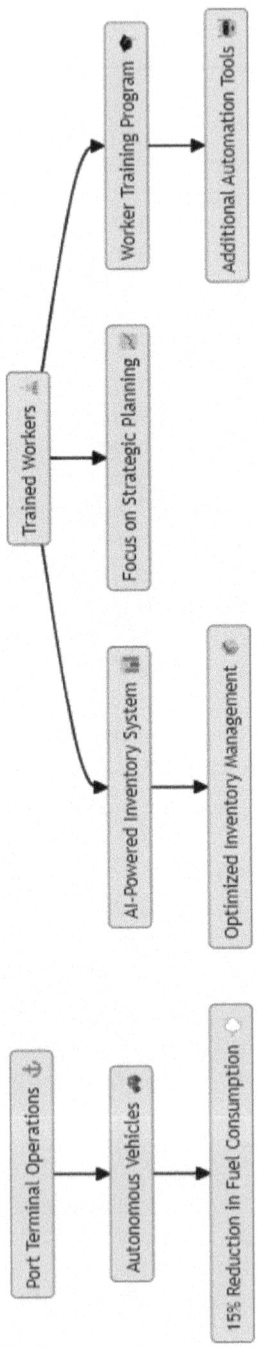

Fig. 7.3 Flowchart illustrating Maersk's human-centric automation of port processes (*Source*: Authors).

Furthermore, ML models are being employed to learn from operational data and progressively improve processes over time, which results in cost savings, enhanced resource utilisation, and operational efficiency. Through these systems, Maersk ensures a data-driven approach that promotes sustainability while also improving responsiveness to market changes, thereby establishing a stronger and more reliable logistics framework (Dagar et al., 2024).

7.4.3.4 Worker Empowerment and the Role of Cobots

A key feature of Maersk's human-centric automation is the deployment of collaborative robots (cobots), designed to work alongside human employees rather than replace them. These cobots assist in physically demanding or repetitive tasks, such as handling and moving smaller containers, which reduces the physical burden on workers and improves workplace safety. By taking over mundane tasks, cobots free up workers to focus on more complex, decision-making roles, such as quality assessment and strategic planning in logistics operations (Maersk, 2022; Dagar et al., 2024).

This integration of cobots highlights Maersk's commitment to Industry 5.0's emphasis on collaboration between humans and intelligent systems. The cobots are programmed to work interactively and intuitively, making the workspace more flexible and accommodating for the human workforce. This collaborative approach has not only improved productivity but also positively impacted employee morale by removing the most taxing aspects of their work while maintaining a sense of job security and meaningful involvement (Maersk, 2022; Dagar et al., 2024).

The effect of worker empowerment can be measured in terms of both productivity and worker satisfaction. Surveys conducted within Maersk reveal a notable increase in employee engagement and a significant reduction in workplace incidents since the introduction of automation and cobot collaboration. These metrics underscore the dual impact of technological integration: while operational efficiency improves, human well-being is also prioritised (Dagar et al., 2024).

7.4.3.5 AI-Driven Sustainability Monitoring

Maersk's use of AI extends to the monitoring of sustainability metrics, ensuring compliance with CE and environmental goals. AI systems continuously monitor emissions, fuel efficiency, and the condition of port terminal equipment, alerting human operators when there is a deviation from expected performance. This proactive monitoring supports preventative maintenance—reducing downtimes and ensuring that all equipment functions within optimal parameters.

Additionally, AI aids in calculating the carbon footprint of each operation, providing transparency into the environmental impact of different routes, modes of transport, and logistic strategies. With this information, Maersk can make more sustainable decisions, such as optimising routes to reduce fuel consumption or shifting more operations to greener energy sources. The AI systems, thus, provide

Maersk with the tools to act on sustainability commitments with precision and agility (Maersk, 2022; Dagar et al., 2024).

7.4.3.6 The Future of Human-Centric Logistics at Maersk

Maersk's human-centric approach to automation is indicative of a broader industry trend towards integrating advanced technologies in a way that complements human labour rather than replaces it. Looking ahead, Maersk aims to expand the role of AI and cobots to other parts of the supply chain, including warehousing and last-mile delivery, to further improve efficiency and sustainability (Maersk, 2022).

The focus will be on ensuring that the introduction of new technologies is done in tandem with upskilling initiatives for employees, thereby equipping them with the knowledge required to work effectively alongside these advanced systems. As logistics continue to evolve, the ability of companies like Maersk to balance automation with human empowerment will set a benchmark for sustainable and resilient logistics practices across the industry (Maersk, 2022; Dagar et al., 2024).

Figure 7.4 presents a structured visualisation of the interactions and roles within Maersk's logistics automation framework, highlighting the collaborative relationship between human operators and various automated systems. The "Worker" class is central, engaging with key components such as Training Programs, Autonomous Vehicles, the AI Inventory System, and other Automation Tools. Workers first participate in training programs to gain proficiency with these automation tools, enabling them to effectively monitor and oversee the operations of autonomous vehicles. They also interact with the AI Inventory System, requesting and receiving real-time data to support informed decision-making. Throughout the process, autonomous vehicles continuously provide operational updates, while the AI system ensures that workers have timely access to crucial information. This diagram effectively illustrates the seamless integration between human roles and automated technologies, showcasing how Maersk leverages these interactions to enhance the efficiency and resilience of its logistics operations.

By prioritising both human-centric and environmentally conscious solutions, Maersk demonstrates how Industry 5.0 principles can be applied to logistics operations to create a balance between technological innovation and human involvement, ultimately leading to a more sustainable, resilient, and adaptive supply chain.

7.4.3.7 Practical Implications of Human-Centric Automation at Maersk

Maersk's implementation of human-centric automation has significant practical implications for the logistics industry, showcasing a progressive model that combines technological advancement with workforce empowerment. The integration of autonomous vehicles, AI systems, and collaborative robots demonstrates that automation can effectively coexist with a skilled human workforce. By leveraging automation for tasks that are repetitive or physically demanding, Maersk not only enhances operational efficiency and reduces environmental impact but also ensures

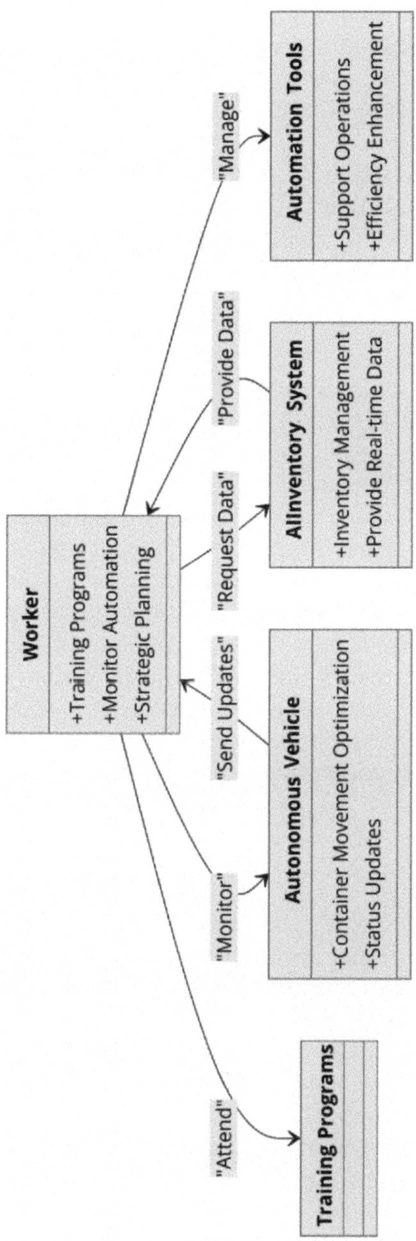

Fig. 7.4 Class diagram showing employee operations alongside automated systems (*Source:* Authors).

that employees remain at the core of its processes. Workers are empowered to make strategic decisions and receive targeted training to interact seamlessly with automation, fostering an adaptable and resilient workforce. This human-centric model leads to increased worker satisfaction, improved safety, and optimised logistics performance, aligning with both Industry 5.0 and circular economy objectives. Ultimately, Maersk's approach serves as a benchmark for balancing technological innovation with human value, offering a sustainable and scalable blueprint for the logistics sector that enhances both productivity and employee well-being (Maersk, 2022; Dagar et al., 2024).

7.4.4 Patagonia's Closed-Loop Logistics Network

Patagonia, renowned for its steadfast commitment to environmental sustainability, has developed a sophisticated closed-loop logistics network that exemplifies the principles of both the circular economy (CE) and Industry 5.0. This approach is a comprehensive system that encourages customers to return used products, which are subsequently repaired, resold, or recycled based on their condition. This process reduces waste, extends product lifespans, and supports CE's core objective of maximising resource efficiency and minimising environmental footprint (Beckman et al., 2019).

The closed-loop model allows Patagonia to address critical issues within the apparel industry, particularly those linked to linear production and waste accumulation. Through this approach, Patagonia effectively reduces the need for new raw materials, thereby cutting down the carbon footprint associated with manufacturing. By integrating human expertise and advanced technology, Patagonia not only enhances operational resilience but also creates a system in which sustainability is intrinsic to every stage of the product life cycle (Patagonia, 2009).

7.4.4.1 Human-Centric Repair and Customer Engagement

A pivotal element of Patagonia's closed-loop logistics network is its emphasis on human-centric repair and refurbishment processes, bolstered by the use of Industry 5.0 technologies. At the forefront of this effort is artificial intelligence (AI), which evaluates the condition of returned products and determines whether they should be repaired, refurbished, or recycled. This allows Patagonia to allocate resources efficiently, ensuring that valuable human labour is directed towards activities where it can add the most value (Beckman et al., 2019).

Once AI-driven assessments are complete, collaborative robots (cobots) work alongside human technicians to carry out the necessary repairs. Cobots are particularly useful for repetitive, physically demanding tasks, such as sewing, quality checks, or other parts of the repair process, which enhances efficiency and reduces the risk of injuries. The involvement of skilled human workers ensures that the integrity and high quality of Patagonia's products are maintained. This blend of technology and human labour optimises the repair process, ultimately extending the useful life of products and supporting the company's sustainability goals (Beckman et al., 2019).

Patagonia involves customers in its closed-loop system through the "Worn Wear" program. This initiative encourages customers to return their used garments in exchange for store credit, transforming them into active contributors to the sustainability process. This collaboration fosters a culture of shared responsibility, aligning with Industry 5.0's emphasis on human-centricity by actively involving consumers in Patagonia's CE practices. Customers not only gain economic value through store credits but also an enhanced sense of participation in environmental stewardship. Through this active customer engagement, Patagonia builds a community dedicated to sustainable consumption, strengthening its brand loyalty while reinforcing responsible consumer behaviour. Figure 7.5 highlights Patagonia's closed-loop logistics process, emphasising how customer returns are integrated into a sustainable recycling, refurbishment, or resale workflow.

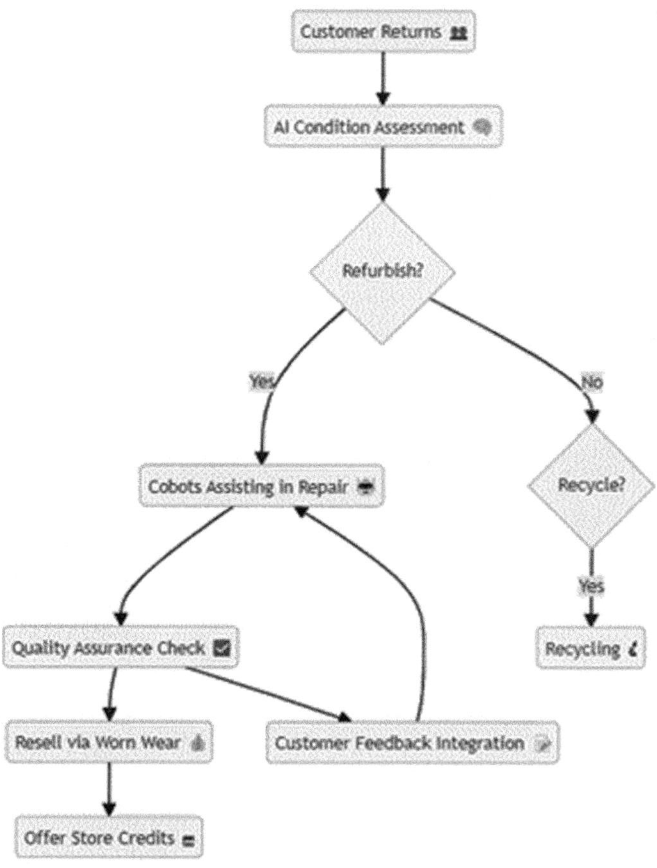

Fig. 7.5 Patagonia's closed-loop process in logistics.

7.4.4.2 AI-Driven Insights and Continuous Improvement

The integration of AI within Patagonia's closed-loop logistics network goes beyond product assessment and repair management. AI also plays a vital role in driving continuous improvement across Patagonia's operations, thereby aligning with Industry 5.0 principles. By collecting and analysing data from returned items, AI systems provide insights into product performance, customer preferences, and common issues that lead to returns. This data-driven feedback loop allows Patagonia to proactively address design flaws, enhance product durability, and even explore innovative materials that can extend the lifespan of their garments (Beckman et al., 2019).

These AI-driven insights support a broader, adaptive approach to product design and manufacturing. For example, if repeated patterns of wear or damage are identified in specific products, Patagonia's design team can use this information to make targeted improvements, thereby reducing future returns and enhancing the overall quality of their offerings. This continuous improvement process helps to close the loop more effectively, minimising waste and reducing the environmental impact of their products (Beckman et al., 2019).

Moreover, AI also aids in inventory management by predicting the availability of returned items for resale and refurbishment. This predictive capacity enhances Patagonia's ability to balance inventory, reducing the need for overproduction and thereby cutting down on resource use and waste. It also optimises the logistics of the "Worn Wear" program, ensuring that refurbished products reach the market efficiently and in response to actual demand rather than speculative production (Beckman et al., 2019).

7.4.4.3 Customer Education and Industry Collaboration

Patagonia's commitment to the circular economy extends beyond its logistics network; it actively focuses on customer education and industry collaboration to maximise the impact of its sustainability initiatives. The company's closed-loop system relies heavily on informed customers who understand the value of returning used products and the process of recycling or refurbishing them. To this end, Patagonia runs awareness campaigns and workshops that educate consumers about the environmental benefits of reusing, recycling, and minimising consumption. These educational initiatives are not merely promotional tools; they represent a vital part of Patagonia's strategy to instill sustainable practices across its customer base, thus enhancing the effectiveness of the closed-loop model (Beckman et al., 2019).

In addition, Patagonia collaborates extensively with other companies, NGOs, and research institutions to tackle broader challenges associated with single-use packaging and textile recycling. For example, Patagonia is part of industry-wide coalitions aimed at innovating new materials for packaging that reduce reliance on plastic and are more easily recyclable. These efforts help Patagonia build more resilient supply chains, enhance recycling capabilities, and influence industry standards towards adopting more circular practices. Such collaborations amplify Patagonia's efforts, as the shared knowledge and collective actions drive systemic change beyond what any single company could achieve alone (Patagonia, 2009; Beckman et al., 2019).

Patagonia's partnerships are also instrumental in expanding the closed-loop logistics concept to new areas, such as packaging and distribution. By working with material suppliers to develop alternatives to conventional polybags and other types of single-use packaging, Patagonia demonstrates its commitment to holistic sustainability. These efforts aim not only to reduce waste within Patagonia's own supply chain but also to encourage the adoption of more sustainable practices industry-wide (Beckman et al., 2019).

7.4.4.4 Practical Implications of Patagonia's Closed-Loop Logistics

The practical implications of Patagonia's closed-loop logistics network are profound, with significant benefits for sustainability, economic resilience, and customer engagement. By integrating human workers, AI, and cobots, Patagonia has managed to create an effective synergy that maximises operational efficiency while maintaining a focus on human empowerment. The AI-driven product assessments and cobot-assisted repairs ensure that returned items are processed quickly and efficiently, thus enabling Patagonia to minimise waste and maximise the utility of each product (Beckman et al., 2019).

From an economic perspective, Patagonia's approach also provides an effective hedge against raw material shortages and supply chain disruptions. By relying less on virgin materials and focusing more on recycling and refurbishment, Patagonia insulates itself from volatile material costs and supply issues. This strategy not only makes the company more resilient but also aligns with its commitment to minimising its environmental footprint. Furthermore, the data gathered from returned products provides valuable insights that inform future product designs, ultimately reducing production inefficiencies and enhancing the overall quality of Patagonia's offerings (Beckman et al., 2019).

Another key implication of Patagonia's closed-loop logistics is the empowerment of consumers as active participants in sustainability efforts. Programs like "Worn Wear" foster deeper brand loyalty and encourage responsible consumer behaviour, which helps to extend the impact of Patagonia's sustainability practices beyond the confines of its direct operations. By educating customers and involving them in the lifecycle of products, Patagonia not only ensures a steady stream of materials for its closed-loop system but also contributes to the cultural shift towards sustainable consumption (Beckman et al., 2019).

Lastly, Patagonia's collaborative efforts within the industry set an important precedent for shared sustainability. By partnering with other companies and research organisations, Patagonia has amplified its impact, helping to establish broader networks of circular practices that benefit the industry as a whole. This collective approach to solving systemic issues—such as single-use packaging waste—illustrates how businesses can lead the way in promoting more sustainable, resilient, and inclusive economic models.

In conclusion, Patagonia's closed-loop logistics network is a comprehensive implementation of CE and Industry 5.0 principles. It exemplifies how advanced technology, human labour, and customer engagement can be integrated to create a sustainable, resilient, and efficient supply chain. Patagonia's emphasis on human-centric repair processes, AI-driven continuous improvement, and industry collaboration not only supports its sustainability objectives but also strengthens its brand, reduces costs, and fosters a deeper connection with customers. As a result, Patagonia's model serves as a benchmark for other industries seeking to adopt circular and human-centered approaches to their supply chain and sustainability practices.

7.4.5 Summary of Case Studies

These case studies illustrate how the integration of Industry 5.0 and CE principles can transform logistics operations for greater sustainability and resilience. By emphasising human-centricity, collaboration, and the use of advanced technologies, companies like DHL, IKEA, Maersk, and Patagonia have demonstrated the potential for logistics to contribute to a circular and sustainable economy. The examples provided emphasise the importance of combining human expertise with technological innovation to create adaptive and efficient logistics systems that minimise waste, optimise resource use, and enhance the well-being of workers. The next section explores the challenges and opportunities associated with implementing Industry 5.0 and CE practices in logistics, providing insights into how companies can overcome barriers to adoption and maximise the benefits of these transformative approaches.

7.5 Challenges and Opportunities in Implementing Circular Economy and Industry 5.0 in Logistics

The integration of Circular Economy (CE) principles and Industry 5.0 in logistics presents numerous opportunities to enhance sustainability, efficiency, and resilience. However, it also introduces several challenges that organisations must navigate to achieve successful implementation. This section explores the key challenges and opportunities associated with adopting CE and Industry 5.0 practices in logistics, focusing on technology, organisational culture, stakeholder engagement, and regulatory considerations.

7.5.1 Technological Challenges and Opportunities

One of the most significant challenges in implementing CE and Industry 5.0 in logistics is the need for advanced technologies that enable greater efficiency and sustainability. Technologies such as artificial intelligence (AI), blockchain, and collaborative robotics are essential for optimising resource use, enabling transparency across supply chains, and enhancing collaboration between humans and machines (Latha & Patil, 2024; Nahavandi, 2019). However, adopting these technologies requires substantial investments in infrastructure and expertise, which may be prohibitive for some organisations, especially small and medium-sized enterprises (SMEs). Limited access to data and insufficient data processing capabilities are common challenges that hinder the full utilisation of these advanced technologies (Gayda, 2023; Vasiliauskas et al., 2024).

Despite these challenges, the technological advancements associated with Industry 5.0 offer significant opportunities for the logistics sector. For example, AI can enhance decision-making processes by providing real-time insights into inventory management, transportation routes, and energy use (Dabo & Hosseinian-Far, 2023). This enables logistics managers to make informed decisions that minimise resource consumption and waste, aligning with CE principles (Chauhan et al.,2022). Additionally, blockchain technology provides an opportunity to create transparent and secure supply chains that track products and materials throughout their lifecycle, thereby supporting the implementation of closed-loop systems (Saberi et al., 2019; Ivanov & Dolgui, 2023).

Collaborative robots (cobots) are another technological opportunity that facilitates the integration of CE and Industry 5.0 in logistics. Cobots work alongside human employees, automating repetitive tasks and reducing physical strain while allowing workers to focus on more complex, value-added activities. This not only enhances operational efficiency but also improves the safety and well-being of employees, which is a core aspect of Industry 5.0's human-centric approach (Nahavandi, 2019; Rahman et al., 2024).

7.5.2 Organisational Culture and Skills

Implementing CE and Industry 5.0 practices in logistics also requires a shift in organisational culture and an investment in skills development. The adoption of advanced technologies and circular practices necessitates new ways of thinking and working, which can be challenging for organisations that are accustomed to linear, efficiency-driven models. Resistance to change is a common barrier, particularly in industries where traditional practices have been well-established over time (Gomes, 2020; Yaqot, 2024). The cultural shift necessary for adopting CE and Industry 5.0 requires an emphasis on innovation, flexibility, and openness to new ways of operation, which might be difficult for organisations with entrenched practices (Gayda, 2023; Martini et al., 2024).

To overcome this challenge, organisations need to foster a culture that embraces innovation, sustainability, and collaboration (Dabo & Hosseinian-Far, 2023). This involves not only investing in new technologies but also in training and development programs that equip employees with the skills needed to work with these technologies effectively (Dagar et al., 2024). Skills such as data analysis, systems thinking, and cross-functional collaboration are crucial for successfully implementing CE and Industry 5.0 practices. By empowering employees with these skills, organisations can facilitate a smoother transition to more sustainable and human-centric logistics operations (Raffik et al., 2023).

In addition to skills development, fostering a culture of collaboration is essential for achieving the goals of CE and Industry 5.0. Logistics operations are inherently complex and involve multiple stakeholders, including suppliers, customers, and regulatory bodies. Organisations that prioritise collaboration and knowledge sharing are better positioned to create adaptive and resilient supply chains that support CE principles. This is particularly important in the context of reverse logistics and closed-loop systems, where effective collaboration is key to ensuring the seamless flow of products and materials (Dabo & Hosseinian-Far, 2023; Vasiliauskas et al., 2024).

7.5.3 Stakeholder Engagement and Collaboration

The success of CE and Industry 5.0 in logistics also depends on effective stakeholder engagement. Stakeholders, including suppliers, customers, regulators, and employees, all play a role in determining the sustainability of logistics operations. Engaging these stakeholders in the transition to CE and Industry 5.0 is essential for ensuring buy-in and support for new practices and technologies (Gomes, 2020; Saberi et al., 2019).

One of the challenges in stakeholder engagement is the need for alignment among diverse groups with varying interests and priorities. For instance, suppliers may be hesitant to adopt circular practices if they perceive them as costly or disruptive, while customers may not be aware of the benefits of circular products and services (Zhang et al., 2023). To address this challenge, organisations must actively communicate the value of CE and Industry 5.0 to stakeholders, highlighting both the environmental and economic benefits (Narula et al., 2023). This can be achieved through education and awareness campaigns, as well as through incentives that encourage stakeholders to adopt more sustainable practices (Saberi et al., 2019).

Collaboration with external stakeholders, such as government bodies and industry associations, also presents an opportunity to advance the adoption of CE and Industry 5.0 in logistics. Government regulations and policy frameworks can play a crucial role in encouraging or mandating the adoption of sustainable practices, providing both a challenge and an opportunity for organisations (Gayda, 2023; Dabo & Hosseinian-Far, 2023). On one hand, regulatory requirements can be complex and costly to comply with; on the other hand, they can create a level playing field and drive industry-wide adoption of circular and sustainable practices. Public-private partnerships and collaborations with industry associations can also facilitate knowledge sharing and innovation, helping organisations overcome barriers to the implementation of CE and Industry 5.0 (Vasiliauskas et al., 2024).

7.5.4 Financial Considerations

The financial implications of implementing CE and Industry 5.0 practices in logistics are a significant challenge for many organisations. The transition to more sustainable and human-centric logistics systems often requires substantial upfront investments in new technologies, infrastructure, and employee training. For SMEs in particular, these costs can be a major barrier to adoption (Latha & Patil, 2024; Beckman et al., 2019). However, there are also significant financial opportunities associated with CE and Industry 5.0. For example, adopting circular practices such as resource optimisation, waste reduction, and reverse logistics can lead to cost savings in the long term. By designing products and supply chains for circularity, organisations can reduce material costs, minimise waste disposal fees, and generate new revenue streams from refurbished or recycled products. Industry 5.0's emphasis on human-centric approaches also leads to opportunities for reducing operational costs by improving employee productivity, satisfaction, and safety (Nahavandi, 2019).

To mitigate the financial challenges associated with the transition to CE and Industry 5.0, organisations can explore a variety of funding and financing options. Government grants, subsidies, and incentives for sustainable practices are available in many regions and can help offset the costs of technology adoption and skills development (Gomes, 2020). Additionally, organisations can partner with other companies, research institutions, or NGOs to share costs and resources, thereby reducing the financial burden of transitioning to more sustainable logistics systems (Dabo & Hosseinian-Far, 2023; Beckman et al., 2019).

7.5.5 Supply Chain Resilience

Implementing CE and Industry 5.0 practices can enhance the resilience of logistics and supply chain operations, but doing so also presents challenges related to supply chain complexity and risk management. Closed-loop supply chains and circular logistics systems often require more intricate planning and coordination compared to traditional linear supply chains. This complexity can lead to increased risk, particularly in terms of supply chain disruptions, inventory management, and the need for new supplier relationships (Dabo & Hosseinian-Far, 2023; Shin et al., 2024).

Nevertheless, CE and Industry 5.0 offer opportunities to improve supply chain resilience by fostering adaptability, diversification, and transparency (Ivanov, 2023). Circular practices such as remanufacturing, refurbishing, and recycling help reduce dependence on virgin materials, thereby mitigating risks associated with resource scarcity and supply chain disruptions. Blockchain technology, which is often associated with Industry 5.0, provides an opportunity to enhance transparency and traceability across the supply chain, enabling organisations to quickly identify and respond to potential risks (Saberi et al., 2019). Moreover, the human-centric approach of Industry 5.0 emphasises collaboration and adaptability, which are critical for navigating the complexities of circular supply chains and ensuring business continuity in the face of disruptions (Zhang et al.,2023).

7.5.6 *Summary of Challenges and Opportunities*

The integration of Circular Economy principles and Industry 5.0 in logistics presents both challenges and opportunities for organisations. Key challenges include the need for technological investments, cultural shifts, stakeholder engagement, financial considerations, and managing supply chain complexity. However, the opportunities associated with CE and Industry 5.0—such as enhanced efficiency, reduced costs, improved supply chain resilience, and a more satisfied and empowered workforce—make the effort worthwhile.

By adopting a proactive approach to technology adoption, skills development, stakeholder collaboration, and financial planning, organisations can successfully overcome these challenges and unlock the full potential of CE and Industry 5.0 in logistics. The next section will provide recommendations for companies looking to implement these practices, focusing on best practices, strategies for overcoming barriers, and the steps required to build a more sustainable and human-centric logistics system.

7.6 Recommendations for Implementing Circular Economy and Industry 5.0 in Logistics

The successful integration of Circular Economy (CE) principles and Industry 5.0 practices into logistics requires careful planning, strategic investments, and an inclusive approach that involves all stakeholders. This section provides actionable recommendations for organisations seeking to implement CE and Industry 5.0 practices, focusing on the key areas of technology adoption, organisational change, stakeholder engagement, financial strategies, and supply chain management. These recommendations aim to help logistics managers and practitioners overcome common barriers and maximise the potential benefits of a sustainable and human-centric logistics system.

7.6.1 *Developing a Clear Strategy and Vision*

Organisations seeking to implement CE and Industry 5.0 practices in logistics must start by developing a clear strategy and vision that outlines their goals, objectives, and expected outcomes. A well-defined strategy ensures alignment with broader business goals and highlights how CE and Industry 5.0 can support long-term value creation. This strategic alignment helps in communicating the purpose of the transition to employees, stakeholders, and partners, ensuring that everyone understands the benefits and is motivated to contribute to the process (Kannan et al., 2024).

To develop an effective strategy, organisations should conduct a thorough assessment of their current logistics operations, identifying areas where CE and Industry 5.0 practices can be integrated to improve efficiency and sustainability. This assessment should evaluate resource use, waste generation, technology adoption, and supply chain resilience. Identifying specific areas for improvement allows organisations to set clear targets and develop a roadmap for implementation, covering both short-term and long-term goals. Such assessments are critical for the successful integration of CE practices into logistics (Payer et al., 2024).

7.6.2 Investing in Technology and Skills Development

The adoption of advanced technologies such as AI, blockchain, and collaborative robots is critical for implementing CE and Industry 5.0 in logistics. Organisations must be willing to invest in these technologies, as well as in the infrastructure required to support them. This may include upgrading existing systems, deploying sensors and data collection devices, and developing integrated digital platforms that enable real-time monitoring and optimisation of logistics operations (Miranda et al., 2015).

Equally important is the investment in skills development to ensure that employees are prepared to work with new technologies and adopt new ways of thinking. Organisations should provide training programs that focus on key skills such as data analysis, systems thinking, digital literacy, and collaboration. Additionally, fostering a culture of continuous learning is essential to keep up with rapid technological advancements and evolving industry standards (Arshad et al., 2017; Nahavandi, 2019). By empowering employees with the skills they need, organisations can create a workforce that is agile, adaptable, and capable of driving innovation.

7.6.3 Fostering a Culture of Innovation and Collaboration

The transition to CE and Industry 5.0 requires a cultural shift that encourages innovation, collaboration, and a commitment to sustainability. Organisations should promote an open and inclusive culture where employees are encouraged to share ideas, experiment with new approaches, and contribute to sustainability initiatives. This can be achieved through innovation labs, collaborative workshops, and employee engagement programs that focus on CE and Industry 5.0 (De Donno et al., 2019).

Collaboration is not only important within the organisation but also with external stakeholders, including suppliers, customers, research institutions, and industry associations. Engaging in partnerships and knowledge-sharing initiatives can help organisations access new ideas, technologies, and best practices that support the implementation of CE and Industry 5.0. For example, collaborating with suppliers can help identify opportunities for reducing waste and optimising resource use, while partnerships with research institutions can facilitate the development of innovative logistics solutions (Askoxylakis, 2018).

7.6.4 Engaging Stakeholders in the Transition

Engaging stakeholders is critical for the successful implementation of CE and Industry 5.0 in logistics. This includes not only internal stakeholders, such as employees and management, but also external stakeholders, such as suppliers, customers, regulatory bodies, and community organisations. Organisations should actively communicate the benefits of CE and Industry 5.0 to stakeholders, using clear and compelling messages that emphasise both the environmental and economic value of these practices (Miraz et al., 2015).

One effective way to engage stakeholders is through collaborative initiatives, such as joint projects, workshops, or pilot programs, that allow stakeholders to experience the benefits of CE and Industry 5.0 firsthand. Additionally, organisations can create incentive programs that encourage suppliers and customers to adopt more sustainable practices. For example, offering preferential contracts to suppliers who demonstrate a commitment to CE or providing discounts to customers who return products for recycling can help drive the adoption of circular practices across the value chain (Miraz et al., 2015).

7.6.5 Leveraging Financial Support and Incentives

Financial considerations are often a major barrier to the adoption of CE and Industry 5.0 practices, particularly for SMEs. To overcome this challenge, organisations should explore various funding and financing options that can help offset the costs of technology adoption, skills development, and infrastructure upgrades. Government grants, subsidies, and tax incentives for sustainability initiatives are available in many regions and can provide valuable financial support for organisations transitioning to CE and Industry 5.0 (European Commission, 2019).

In addition to government support, organisations can seek partnerships with private investors, venture capital firms, or industry associations that are interested in supporting sustainable innovation. Green finance initiatives, such as sustainability-linked loans or bonds, are also becoming increasingly popular and can provide organisations with the capital they need to invest in CE and Industry 5.0 (Nahavandi, 2019). By leveraging these financial opportunities, organisations can reduce the upfront costs of implementing sustainable and human-centric logistics systems.

7.6.6 Building Resilient and Circular Supply Chains

The successful implementation of CE and Industry 5.0 in logistics also requires a focus on building resilient and circular supply chains. Organisations should design their supply chains with circularity in mind, focusing on reducing waste, extending product life cycles, and enabling the efficient recovery of resources. This may involve redesigning products for easier disassembly and recycling, establishing reverse logistics systems, and collaborating with suppliers to ensure that materials can be recovered and reused (Albreem et al., 2021).

Blockchain technology can be leveraged to enhance transparency and traceability across the supply chain, providing real-time information about the movement of products and materials. This can help organisations identify potential risks, such as supply chain disruptions or resource shortages, and take proactive measures to address them (Fernández-Caramés & Fraga-Lamas, 2019). Additionally, the adoption of predictive analytics and AI can help optimize inventory management, reduce waste, and improve supply chain resilience.

Building resilient supply chains also involves diversifying suppliers and reducing dependence on single sources of critical materials. By working with a network of suppliers and developing alternative sourcing strategies, organisations can reduce their exposure to risks and ensure the continuity of their logistics operations in the face of disruptions (Miraz et al., 2015).

7.6.7 Measuring and Communicating Impact

To demonstrate the value of CE and Industry 5.0 practices, organisations must establish clear metrics for measuring the impact of their initiatives. Key performance indicators (KPIs) should be developed to track progress in areas such as resource efficiency, waste reduction, carbon emissions, employee well-being, and supply chain resilience (Bouchery et al., 2010). By monitoring these metrics, organisations can assess the effectiveness of their CE and Industry 5.0 initiatives and make data-driven decisions to improve their logistics operations.

Communicating the impact of CE and Industry 5.0 initiatives is also important for building trust and credibility with stakeholders. Organisations should share the results of their sustainability efforts through sustainability reports, case studies, and other communication channels. This not only helps demonstrate the value of CE and Industry 5.0 to stakeholders but also enhances the organisation's reputation and brand image as a leader in sustainable and human-centric logistics (Bouchery et al., 2010).

7.6.8 Pilot Programs and Incremental Implementation

Implementing CE and Industry 5.0 practices in logistics can be a complex and resource-intensive process, and organisations may face challenges in scaling up new initiatives. To address this, it is recommended that organisations start with pilot programs that allow them to test new technologies, processes, and collaboration models on a smaller scale before full-scale implementation. Pilot programs provide valuable insights into potential challenges and help organisations refine their approach based on real-world experience (Banbury et al., 2021).

Incremental implementation is also a practical approach for managing the transition to CE and Industry 5.0. By breaking down the transition into smaller, manageable phases, organisations can focus on achieving specific milestones and gradually build the capabilities needed for full implementation (Miranda et al., 2015). This approach also allows organisations to allocate resources more effectively and minimise disruptions to existing logistics operations.

7.6.9 Summary of Recommendations

The integration of Circular Economy principles and Industry 5.0 in logistics presents both challenges and opportunities. The recommendations outlined in this section provide a framework for organisations seeking to implement these practices, focusing on developing a clear strategy, investing in technology and skills development, fostering a culture of innovation and collaboration, engaging stakeholders, leveraging financial support, building resilient and circular supply chains, measuring impact, and adopting a phased implementation approach.

By following these recommendations, organisations can overcome the challenges associated with CE and Industry 5.0 and unlock the full potential of a sustainable, efficient, and human-centric logistics system. The next section will conclude the chapter by summarising the key insights and outlining the future prospects for CE and Industry 5.0 in logistics.

7.7 Conclusion

The integration of Circular Economy (CE) principles and Industry 5.0 practices in logistics presents a promising path towards creating a more sustainable, efficient, and human-centric logistics system. This chapter has explored how CE, characterised by reducing waste, promoting resource efficiency, and extending product life cycles, can be synergised with Industry 5.0's emphasis on collaboration between humans and advanced technologies. Together, these paradigms can address some of the most pressing challenges facing the logistics sector today, including environmental impact, resource constraints, supply chain vulnerabilities, and the need for greater adaptability in the face of global disruptions.

The evolution of logistics from Industry 4.0 to Industry 5.0 marks a significant shift towards focusing on the well-being of people and the environment, emphasising the importance of collaboration between technology and human ingenuity. By leveraging technologies such as AI, robotics, blockchain, and advanced analytics, logistics operations can become smarter and more responsive, while also reducing their environmental footprint. The adoption of CE practices ensures that logistics processes contribute to a regenerative economic model that prioritises resource recovery, minimises waste, and creates a positive environmental impact.

One of the key insights from this chapter is that the successful implementation of CE and Industry 5.0 in logistics requires more than just technological advancements; it also necessitates a cultural shift that promotes innovation, collaboration, and a commitment to sustainability. Organisations must foster an environment where employees are empowered to contribute to sustainability initiatives, share ideas, and leverage new technologies to drive positive change. By focusing on skills development and creating an inclusive culture, logistics organisations can ensure that their workforce is well-prepared to adapt to new technologies and embrace sustainable practices.

Stakeholder engagement is also a critical component of the transition to CE and Industry 5.0. Effective collaboration with suppliers, customers, regulatory bodies, and industry associations can facilitate the adoption of circular practices and drive systemic change across the entire value chain. Engaging stakeholders in the development of sustainable logistics solutions helps build trust, enhances the organisation's reputation, and creates opportunities for innovative partnerships that can generate shared value.

The recommendations provided in this chapter serve as a practical guide for organisations seeking to implement CE and Industry 5.0 practices in logistics. These recommendations emphasise the importance of developing a clear strategy and vision, investing in technology and skills development, fostering a culture of innovation, engaging stakeholders, leveraging financial support, building resilient supply chains, and adopting a phased implementation approach. By following these recommendations, organisations can effectively navigate the challenges of transitioning to a sustainable logistics system and unlock the full potential of CE and Industry 5.0.

Looking to the future, the logistics sector is likely to continue evolving as new technologies and practices emerge. The convergence of CE and Industry 5.0 represents a significant opportunity to redefine how logistics systems operate, making them more adaptable, efficient, and aligned with the principles of sustainability. As the world faces increasing environmental challenges, such as climate change, resource depletion, and biodiversity loss, the need for sustainable logistics solutions has never been greater. CE and Industry 5.0 provide a blueprint for addressing these challenges, enabling organisations to create logistics systems that are not only economically viable but also environmentally and socially responsible.

In conclusion, the integration of CE and Industry 5.0 in logistics has the potential to transform the sector, making it a driving force for sustainability and human well-being. By adopting a holistic approach that combines technological innovation with a commitment to circularity and human-centered values, logistics organisations can contribute to a more sustainable future while enhancing their competitiveness in the global market. The path to achieving this vision may be complex, but the rewards are substantial—for the environment, for society, and for the long-term success of the logistics industry.

Declaration of Generative AI and AI-assisted Technology in the Writing Process

During the preparation of the manuscript, the authors used specialist referencing software Refworks for reference generation and citation; and SCISPACE's typeset.

io to discover new papers, and get summary, findings, and conclusions from extant literature. After using these tools, the authors reviewed and edited the content as needed and take full responsibility for the content of the paper.

References

Albreem, M.A., Sheikh, A.M., Alsharif, M.H., Jusoh, M. and Mohd Yasin, M.N. (2021). Green Internet of Things (GIoT): Applications, Practices, Awareness, and Challenges. *IEEE Access*, 9, pp. 38833–38858.

Arshad, R., Zahoor, S., Shah, M.A., Wahid, A. and Yu, H. (2017). Green IoT: An investigation on energy-saving practices for 2020 and beyond. *IEEE Access*, 5, pp. 15667–15681.

Askoxylakis, I. (2018). A Framework for Pairing Circular Economy and the Internet of Things. In *Proceedings of the 2018 IEEE International Conference on Communications (ICC)*, Kansas City, MO, USA, 20–24 May 2018, pp. 1–6.

Bag, S., Yadav, G., Wood, L.C., Dhamija, P. and Joshi, S. (2020). Industry 4.0 and the circular economy: Resource melioration in logistics. *Resources Policy*, 68, p. 101776. https://doi.org/10.1016/j.resourpol.2020.101776.

Banbury, C., Reddi, V.J., Torelli, P., Holleman, J., Jeffries, N., Kiraly, C., Montino, P., Kanter, D., Ahmed, S., Pau, D. and Thakker, U. (2021). Mlperf tiny benchmark. *arXiv preprint arXiv:2106.07597.*

Beckman, S., O'Neill, K., Pendleton-Knoll, S., Rosenzweig, W., and Strand, R. (2019). Patagonia: Closing the Loop on Packaging Pollution. *University of California.* Retrieved from https://sustainabilitycasecompetition.pdf.

Bhowmik, O., Chowdhury, S., Ashik, J.H., Mahmud, G.I., Khan, M.M. and Hossain, N.U.I. (2024). Application of artificial intelligence in reverse logistics: A bibliometric and network analysis. *Supply Chain Analytics*, 7, p. 100076. https://doi.org/10.1016/j.sca.2024.100076.

Borumand, A. and Nookabadi, A.S. (2024). Adjustable robust optimization for fleet sizing problem in closed-loop supply chains with simultaneous delivery and pickup. *Optimization Online.* Retrieved from https://optimization-online.org/wp-content/uploads/2024/02/Adjustable-robust-optimization-for-fleet-sizing-problem-in-closed-loop-supply-chains-with-simultaneous-delivery-and-pickup.pdf.

Bouchery, Y., Ghaffari, A. and Jemai, Z. (2010). Key Performance Indicators for Sustainable Distribution Supply Chains: Set Building Methodology and Application. *Proceedings of APMS 2010 - International Conference on Advances in Production Management Systems.*

Bressanelli, G., Adrodegari, F., Pigosso, D.C. and Parida, V. (2022). Towards the smart circular economy paradigm: a definition, conceptualization, and research agenda. *Sustainability*, 14(9), p. 4960. https://doi.org/10.3390/su14094960.

Calon, S.R.J.L. (2020). Circular Design Strategies Enabling Re-use in High-Technology Manufacturing Firms (Doctoral dissertation, MSc Thesis, Eindhoven University of Technology). Retrieved from https://research.tue.nl/files/167328397/Master_Thesis_Stef_Calon.pdf.

Chauhan, C., Parida, V. and Dhir, A. (2022). Linking circular economy and digitalisation technologies: A systematic literature review of past achievements and future promises. *Technological Forecasting and Social Change*, 177, p. 121508. https://doi.org/10.1016/j.techfore.2022.121508.

Chen, Z., Chen, S., Liu, C., Nguyen, L.T. and Hasan, A. (2020). The effects of circular economy on economic growth: A quasi-natural experiment in China. *Journal of Cleaner Production*, 271, p. 122558. https://doi.org/10.1016/j.jclepro.2020.122558.

Dabo, A.A.A. and Hosseinian-Far, A. (2023). An Integrated Methodology for Enhancing Reverse Logistics Flows and Networks in Industry 5.0. *Logistics*, 7(4), p. 97. https://doi.org/10.3390/logistics7040097.

Dagar, M., Tate, M. and Johnstone, D. (2024). Digital Transformation at Maersk: The Never-Ending Pace of Change. *Journal of Information Technology Case and Application Research*, 26(2), pp. 111–143.

De Donno, M., Tange, K. and Dragoni, N. (2019). Foundations and Evolution of Modern Computing Paradigms: Cloud, IoT, Edge, and Fog. *IEEE Access*, 7, pp. 150936–150948.

DHL (2022). Delivering on Circularity: Pathways for Fashion and Consumer Electronics. Deutsche Post DHL Group. Retrieved from https://www.dhl.com/content/dam/dhl/global/core/documents/pdf/DHL_Delivering_on_Circularity_White_Paper_2022.pdf.

DHL (2023a). Sustainable Packaging and Circularity Solutions. DHL Group. Retrieved from https://www.dhl.com/content/dam/dhl/global/csi/documents/pdf/glo-csi-sustainable-packaging-and-circularity.pdf.

DHL (2023b). DIGI x ESG: Digital Path to Sustainability. Presentation from DigiFridays, Deutsche Post DHL Group Investor Relations. Retrieved from https://group.dhl.com/content/dam/deutschepostdhl/de/media-center/investors/documents/capital-markets-days/DPDHL-DigiFridays-Digi-x-ESG-02062023.pdf.

ElMenshawy, O. (2023). Sailing Towards Sustainable Supply Chains: A Multi-Objective Optimization Model for Green Ship Recycling in a Circular Economy (Doctoral dissertation). Dalspace Library. Retrieved from https://dalspace.library.dal.ca/handle/10222/83311.

European Commission. Circular Economy Action Plan. Available online: https://ec.europa.eu/environment/strategy/circular-economy-action-plan_es (accessed on 28 July 2024).

Fernández-Caramés, T.M. and Fraga-Lamas, P. (2018). Design of a fog computing, blockchain and IoT-based continuous glucose monitoring system for crowdsourcing mHealth. In *Proceedings* (Vol. 4, No. 1, p. 37). MDPI.

Fera, M., Caterino, M., Macchiaroli, R. and Pham, D.T. (2024). Advances in Remanufacturing: Proceedings of the VII International Workshop on Autonomous Remanufacturing. Springer Nature. Retrieved from https://link.springer.com/content/pdf/10.1007/978-3-031-52649-7.pdf.

Gayda, S.V. (2023). State and analysis of the dynamics indicators of the production volume of the woodworking and furniture industry. *Forestry, Forest, Paper and Woodworking Industry*, 49, pp. 4–19. Retrieved from https://forest-woodworking.nltu.edu.ua/index.php/journal/article/view/65.

Ghoreishi, M. and Happonen, A. (2020). Key enablers for deploying artificial intelligence for circular economy embracing sustainable product design: three case studies. *AIP Conference Proceedings*. AIP Publishing LLC, Selangor Darul Ehsan, p. 050008.

Gomes, R. (2020). Improving Reverse Logistics Operations: The Case Study of IKEA. *Instituto Superior Técnico.*

Grosse, E.H., Sgarbossa, F., Berlin, C., and Neumann, W.P. (2023). Human-centric production and logistics system design and management: transitioning from Industry 4.0 to Industry 5.0. *International Journal of Production Research*, 61(22), pp. 7749–7759.

Ivanov, D. (2023). The Industry 5.0 framework: viability-based integration of the resilience, sustainability, and human-centricity perspectives. *International Journal of Production Research*, 61(5), pp. 1683–1695. https://doi.org/10.1080/00207543.2022.2118892.

Kannan, D., Amiri, A.S., Shaayesteh, M.T., Nasr, A.K., and Mina, H. (2024). Unveiling barriers to the integration of blockchain-based circular economy and Industry 5.0 in manufacturing industries: A strategic prioritization approach. *Business Strategy and the Environment.*

Latha, C.A. and Patil, M.M. (2024). Artificial Intelligence Applications in Industry 4.0: Applications and Challenges. *AI-Driven Digital Twin and Industry 4.0*, pp. 15–24. Retrieved from https://www.taylorfrancis.com/chapters/edit/10.1201/9781003395416-2.

Maersk (2022). Transforming logistics through technology-enabled digital supply chain solutions. *Maersk.* Retrieved from https://www.maersk.com/news/articles/2022/03/07/transforming-logistics-through-technology-enabled-digital-supply-chain-solutions.

Mahiri, F., Najoua, A., Souda, S.B. and Amini, N. (2023). From Industry 4.0 to Industry 5.0: The transition to human centricity and collaborative hybrid intelligence. *Journal of Hunan University Natural Sciences*, 50(4). https://doi.org/10.55463/issn.1674-2974.50.4.8.

Malmgren, K. and Mötsch Larsson, K. (2020). Reverse logistics in the transition towards circular economy-a case study of customer returns at IKEA.

Martini, B., Bellisario, D. and Coletti, P. (2024). Human-Centered and Sustainable Artificial Intelligence in Industry 5.0: Challenges and Perspectives. *Sustainability*, 16(13), p. 5448.

Mekkunnel, F. (2019). Industry 5.0: Man-Machine Revolution (Doctoral dissertation, Wien).

Miranda, J., Mäkitalo, N., Garcia-Alonso, J., Berrocal, J., Mikkonen, T., Canal, C. and Murillo, J.M. (2015). From the Internet of Things to the Internet of People. *IEEE Internet Computing*, 19(2), pp. 40–47.

Miraz, M.H., Ali, M., Excell, P.S. and Picking, R. (2015). A review on Internet of Things (IoT), Internet of Everything (IoE), and Internet of Nano Things (IoNT). In Proceedings of the 2015 Internet Technologies and Applications.

Mishra, A., Dutta, P., Jayasankar, S., Jain, P. and Mathiyazhagan, K. (2023). A review of reverse logistics and closed-loop supply chains in the perspective of circular economy. *Benchmarking: An International Journal*, 30(3), pp. 975–1020.

Nahavandi, S. (2019). Industry 5.0—A Human-Centric Solution. *Sustainability*, 11(16), p. 4371. https://doi.org/10.3390/su11164371.

Narula, S., Tamvada, J.P., Kumar, A., Puppala, H. and Gupta, N. (2024). Putting digital technologies at the forefront of Industry 5.0 for the implementation of a circular economy in manufacturing industries. *IEEE Transactions on Engineering Management*, 71, pp. 3363–3374. https://doi.org/10.1109/TEM.2023.3344373.

Nguyen, P., Ivanov, D. and Sgarbossa, F. (2023). A digital twin–based approach to reinforce supply chain resilience: simulation of semiconductor shortages. In *IFIP International Conference on Advances in Production Management Systems*, Cham: Springer Nature Switzerland, pp. 563–576.

Owen, J. (2024). Closed-loop supply chain management and industrial symbiosis for SMEs. *ResearchGate.* Retrieved from https://www.researchgate.net/publication/381888470.

Patagonia (2009). Closing the Loop: A report on Patagonia's Common Threads garment recycling program. Retrieved from https://eu.patagonia.com/gb/en/stories/closing-the-loop-a-report-on-patagonias-common-threads-garment-recycling-program/story-19961.html.

Payer, R.C., Quelhas, O.L.G. and Bergiante, N.C.R. (2024). Framework to supporting monitoring the circular economy in the context of industry 5.0: a proposal considering circularity indicators, digital transformation, and sustainability. *Journal of Cleaner Production*, p. 142850.

Pizoń, J., Witczak, M., Gola, A. and Świć, A. (2023). Challenges of Human-Centered Manufacturing in the Aspect of Industry 5.0 Assumptions. *IFAC-PapersOnLine*, 56(2), pp. 156–161.

Rahman, M.M., Khatun, F., Jahan, I., Devnath, R. and Bhuiyan, M.A.A. (2024). Cobotics: The evolving roles and prospects of next-generation collaborative robots in Industry 5.0. *Journal of Robotics*, 2024(1), p. 2918089.

Raffik, R., Sathya, R.R., Vaishali, V. and Balavedhaa, S. (2023). Industry 5.0: Enhancing human-robot collaboration through collaborative robots–A review. In *2023 2nd International Conference on Advancements in Electrical, Electronics, Communication, Computing and Automation (ICAECA)*, IEEE, pp. 1–6.

Saberi, S., Kouhizadeh, M., Sarkis, J. and Shen, L. (2019). Blockchain technology and its relationships to sustainable supply chain management. *International Journal of Production Research*, 57(7), pp. 2117–2135.

Sarkar, B.D., Shardeo, V., Dwivedi, A. and Pamucar, D. (2024). Digital transition from Industry 4.0 to Industry 5.0 in smart manufacturing: A framework for a sustainable future. *Technology in Society*, 78, p. 102649.

Shin, Y., Kim, G. and Jeong, Y. (2024). Robust Closed-Loop Supply Chain Model with Return Management System for Circular Economy. *SSRN*. Retrieved from https://papers.ssrn.com/sol3/papers.cfm?abstract_id=4934197.

Szmelter-Jarosz, A., Abdi, H. and Nozari, H. (2024). Closed-loop supply chain management (CLSCM) in the circular economy. *Journal of Business and Future Economy*, 1(1), pp. 1–24. Retrieved from https://journals.iau.ae/index.php/JBFE/article/view/1.

Tóth, A., Nagy, L., Kennedy, R., Bohuš, B., Abonyi, J. and Ruppert, T. (2023). The human-centric Industry 5.0 collaboration architecture. *MethodsX*, 11, p. 102260. https://doi.org/10.1016/j.mex.2023.102260.

Vasiliauskas, A.V., Ivanauskas, S. and Čižiūnienė, K. (2024). Challenges of ensuring reverse logistics in a military organization using outsourced services. *Sustainability*, 16(11), pp. 1–21. https://doi.org/10.3390/su16114569.

Yaqot, M., Menezes, B., Mohammed, A. and Moloney, K. (2024). A State-of-the-Art Review and Framework for Human-Centric Automation in Industry 5.0. In *IFIP International Conference on Advances in Production Management Systems*, Cham: Springer Nature Switzerland, pp. 385–400.

Zhang, C., Wang, Z., Zhou, G., Chang, F., Ma, D., Jing, Y., Cheng, W., Ding, K. and Zhao, D. (2023). Towards new-generation human-centric smart manufacturing in Industry 5.0: A systematic review. *Advanced Engineering Informatics*, 57, p. 102121. https://doi.org/10.1016/j.aei.2023.102121.

8 Reverse Logistics in Industry 5.0: Opportunities and Challenges: A Systematic Literature Review and Research Agenda

Sina Shahbazi Manshadi[1], Farideh Moosabeiki Dehabadi[2], Andrea Appolloni[3], Xu Sun[4] and Hao Yu[5*]

8.1 Introduction

The evolving dynamics of industrial systems have underscored the importance of integrating sustainability and human-centric approaches in modern manufacturing and logistics processes. This paradigm shift from efficiency-focused systems, as characterized by Industry 4.0, to a more holistic model of Industry 5.0 highlights the need to prioritize environmental, social, and economic well-being alongside technological innovation. The introduction of reverse logistics into this framework further enhances its relevance, especially in light of growing global concerns about resource conservation, waste reduction, and sustainable supply chains (Govindan, 2022).

Reverse logistics, a critical component of circular economy models, involves the systematic recovery, reuse, and recycling of products and materials at the end of their life cycle (Krstić et al., 2022). This process helps reduce environmental impact while enhancing resource optimization and economic viability (Rajput and Singh, 2022). When combined with advanced Industry 5.0 technologies, e.g., artificial intelligence (AI), internet of things (IoT), blockchain, and collaborative robots (cobots), reverse logistics operations are transformed, enabling unprecedented levels of efficiency, accuracy, resilience, transparency, and sustainability (Sun et al., 2022b, Dev et al., 2020, Rajput and Singh, 2022).

The integration of Industry 5.0 into reverse logistics marks a transformative shift from the technology-driven solutions of Industry 4.0 to a balanced interplay of technological capabilities and human values (Sun et al., 2023). By emphasizing sustainability, resilience, and human-centricity, Industry 5.0 technologies not only optimize reverse logistics processes but also align them with the broader goals of environmental stewardship, social well-being, and economic sustainability (Ivanov, 2023). This synthesis creates an opportunity to address critical challenges

[1-3] School of Economics, University of Rome, Tor Vergata, 00133 Rome, Italy.
[4-5] Department of Industrial Engineering, UiT—The Arctic University of Norway, 8514 Narvik, Norway.
* Corresponding author: hao.yu@uit.no

DOI: 10.1201/9781003466451-8

in supply chain and logistics management, particularly in the context of climate change and resource scarcity.

This chapter examines the integration of Industry 5.0 technologies within reverse logistics to evaluate their potential to enhance sustainability across environmental, social, and economic dimensions. The chapter systematically explores six core reverse logistics practices: collection, reuse, repair, recycling, remanufacturing, and disposal, and how Industry 5.0 technologies improve these practices to contribute to the goals of the circular economy is also investigated.

This chapter aims to:

- Provide a comprehensive understanding of the research landscape for reverse logistics within Industry 5.0.
- Understand how Industry 5.0 technologies enhance reverse logistics practices and their sustainability outcomes.
- Identify gaps in current literature and propose future research directions to address challenges in the integration of Industry 5.0 into reverse logistics operations.

A systematic literature review (SLR) forms the methodological foundation of this study, leveraging established frameworks by Denyer & Tranfield 2009. The review identifies, evaluates, and synthesizes academic contributions across domains such as engineering, computer science, and sustainability to analyze the intersection of Industry 5.0 technologies and reverse logistics practices. The SLR focuses on peer-reviewed articles published between 2020 and 2024, drawing on Scopus and Web of Science databases to ensure a robust and unbiased data collection process.

Bibliometric techniques are employed to map the evolution, impact, and structure of research in this area. These techniques provide insights into publication trends, influential works, and thematic clusters, providing a thorough understanding of the field's development. Content analysis complements the bibliometric approach by exploring the specific contributions of Industry 5.0 technologies to reverse logistics practices, emphasizing their role in achieving sustainability goals. The study reveals that Industry 5.0 technologies have significant potential to enhance reverse logistics operations by addressing key challenges, i.e., waste management, resource recovery, and supply chain transparency. Some instances are given as follows:

- IoT and AI optimize collection and recycling processes by enabling real-time tracking, predictive maintenance, and efficient resource allocation.
- Blockchain ensures traceability and accountability across supply chains, facilitating sustainable practices and fostering consumer trust.
- Cobots and digital twins enhance repair and remanufacturing processes by improving precision, reducing costs, and minimizing environmental impact.

Despite these advancements, the study identifies notable gaps in empirical research and practical implementation. For example, human-centric dimensions of Industry 5.0, e.g., worker well-being and ethical supply chain practices, remain underexplored in reverse logistics. Furthermore, technologies like additive manufacturing and immersive technologies (AR/VR) hold untapped potential for advancing sustainable reverse logistics practices.

This chapter contributes to the growing body of knowledge on Industry 5.0 by emphasizing its transformative potential in reverse logistics. By integrating advanced technologies with human-centered principles, Industry 5.0 offers a pathway to more sustainable, resilient, and inclusive supply chain systems. However, realizing this vision requires addressing critical gaps through methodological development, empirical studies, and case-based research to understand the real-world applications and implications of these innovations.

Following the introduction, the rest of the chapter is organized as follows. Section 8.2 provides the theoretical background of Industry 5.0 and the reverse logistics framework. Section 8.3 introduces the research method. Section 8.4 and Section 8.5 provide the bibliometric analysis and content analysis, respectively. The opportunities, challenges, and future research suggestions are discussed in Section 8.6. Finally, Section 8.7 summarizes the paper.

8.2 Industry 5.0 Technologies and Reverse Logistics Framework

The foundation of Industry 5.0 lies in three fundamental principles: putting humans at the center, focusing on sustainability, and building resilience. These three values give substance to the paradigmatic change shifting from the technology-centered approach of Industry 4.0 to a more holistic strategy: the one that is human- and society-centric, prioritizing human needs and societal interests in the industrial process.

Industry 5.0 represents a transformative phase in industrial development, emphasizing cooperation between humans and machines, productivity, and mass customization and individualization through advanced technologies. Unlike Industry 4.0, which focuses on automation and intelligent machines, Industry 5.0 centers on a more human-oriented approach by merging cognitive skills with advanced systems. Several enabling technologies play an essential role in driving this industrial revolution.

- Internet of things (IoT) supports Industry 5.0 by enabling real-time communication and data exchange among humans, machines, and systems. Continuous monitoring of industrial processes enhances predictive maintenance, minimizing downtime and boosting efficiency (Roy Chowdhury et al., 2021). An important usage of IoT in the healthcare sector is in the form of remote patient monitoring, which enables the timely transfer of data obtained from sensors to healthcare providers, thus enabling timely intervention in treatment (Roy Chowdhury et al., 2021).
- Cyber-physical systems (CPS) integrate the physical and virtual worlds, enabling real-time monitoring and control in manufacturing. These systems enhance flexibility and agility in responding to market demands while ensuring

safe human-machine interaction (Wang & Ji, 2018). Advances in IoT, cloud computing, and big data analytics have further evolved CPS into cyber-physical cognitive systems (CPCS), embedding cognitive capabilities like environmental observation and adaptive decision-making. In human-robot collaboration (HRC), CPS allows humans and robots to perform tasks together in real time, enabling a harmonious integration of labor and automation (Chen et al., 2020).

- Artificial intelligence (AI) is a cornerstone of Industry 5.0, enhancing automation and intelligent decision-making. It supports predictive maintenance, process optimization, and quality control by analyzing data collected from IoT devices (Tromp et al., 2020). AI also enables customer personalization by analyzing consumer preferences, facilitating mass customization, and improving efficiency (Choudhary et al., 2020; Siriwardhana et al., 2021).
- Blockchain ensures transparency, safety, and traceability in Industry 5.0 ecosystems. Its decentralized, immutable ledgers promote trust in supply chain management and intellectual property protection (Gamage et al., 2020). Blockchain-based smart contracts streamline automated processes, ensuring secure and efficient operations (Backman et al., 2017).
- Big data analytics in Industry 5.0 will help businesses comprehend how consumers behave when it comes to the pricing of products, enhancing their production efficiency, and reducing overheads accordingly. The difficulty lies in grasping user behavior, social interactions, and human behavioral patterns. Big companies, e.g., Facebook, Twitter, LinkedIn, etc., use big data analytics to enhance products and increase sales concerning consumer satisfaction. Big data has become a significant topic of conversation in both the business world and academic circles (Cheng et al., 2023).
- Additive manufacturing (AM), also referred to as 3D printing, is the backbone of Industry 5.0 for mass customization. Unlike conventional subtractive manufacturing, AM enables the development of highly customized products with less waste and facilitates rapid prototyping and low-volume production (Rani et al., 2024). AM allows industries to fulfill specific customers' demands faster, which is crucial for industries such as aerospace and healthcare due to the need for mass customization.
- Cloud computing manages vast amounts of data, offering scalability and enabling collaborative environments. Meanwhile, edge computing processes data closer to its origin, reducing latency and enhancing real-time decision-making, especially in mission-critical applications like autonomous vehicles and smart factories (Sharma et al., 2024; Siriwardhana et al., 2020). These technologies improve data security, minimize communication overhead, and enable preventive analytics for early issue detection.
- Smart robots and collaborative robots (cobots) are foundational to Industry 5.0's human-centric approach. Cobots enhance productivity by assisting humans in precision-based tasks, ensuring safety through advanced sensors, and reducing physical strain (Wang et al., 2019). Unlike traditional robots,

cobots work alongside humans, facilitating the delivery of customized and personalized products. For example, the Da Vinci surgical system improves surgical outcomes through robotic assistance, showcasing the synergy between humans and technology (Douissard et al., 2019).

- Digital twins replicate physical systems in virtual environments, enabling manufacturers to simulate operations, predict failures, and optimize processes without disrupting actual systems (Tromp et al., 2020). IoT integration makes digital twins more accessible and powerful, allowing real-time updates and customization. These technologies support personalized product development and rapid innovation while reducing maintenance costs (Lu et al., 2020).
- Augmented reality (AR), virtual reality (VR), and extended reality (XR) enhance human-machine interaction by creating immersive environments. These technologies are used in training, enabling workers to learn complex tasks digitally before physical implementation. AR also facilitates remote collaboration by superimposing instructions on real-world objects, improving efficiency and accuracy (Luciano et al., 2005).
- Cybersecurity is essential in Industry 5.0 as connectivity grows, to protect sensitive data and maintain system integrity. Advanced encryption techniques, AI-driven threat detection, and blockchain technologies safeguard critical infrastructure from cyberattacks, ensuring uninterrupted industrial operations (Bakkar, 2023).

Industry 5.0 integrates advanced technologies to create a human-centric, sustainable, and resilient industrial ecosystem, and it also promotes collaboration between humans and machines, enhancing productivity, safety, and innovation. These technologies collectively advance mass customization and individualization, operational efficiency, and sustainability, setting the stage for a more inclusive and environmentally responsible industrial future.

Reverse logistics focuses on managing returned products and materials at the end of their life cycle by reintroducing them into the supply chain for reuse, remanufacturing, recycling, or proper disposal. With the growing emphasis on sustainability and resource optimization in modern industries, reverse logistics has become essential for minimizing waste, reducing environmental impact, and enhancing profitability (Alonso et al., 2021; Govindan, 2022). Fully implemented reverse logistics processes improve efficiency and customer satisfaction while aligning with sustainability goals (Fraga-Lamas et al., 2021; Wiegand and Wynn, 2023). In the Industry 5.0 framework, advanced technologies, i.e., IoT, AI, and big data analytics amplify the efficiency and transparency of reverse logistics practices, enabling companies to better manage product returns and recover value. Reverse logistics consists of six key practices: collection, reuse, refurbishing/repair, recycling, remanufacturing, and disposal (Tavana et al., 2023; Lou et al., 2024).

- Collection is the first step in reverse logistics, involving the retrieval of used products for further processing. IoT and AI technologies enhance this process by tracking product life cycles, predicting returns, and optimizing retrieval schedules (Ma et al., 2022; Salikhov et al., 2023). In the electronics industry, innovative return policies incentivize customers to return old devices, ensuring a steady supply of valuable materials for reuse or recycling (Meier et al., 2023).

- Reuse involves restoring returned goods to the same purpose without significant alteration. This approach reduces raw material consumption and lowers manufacturing costs while minimizing environmental degradation (Govindan, 2022; Al Doghan & Sundram, 2023; Franzè et al., 2023). Industries such as automotive and fashion adopt "take-back" schemes and utilize AI-driven quality assessments to enhance the potential for reuse (Kazancoglu et al., 2023; Charnley et al., 2022).
- Refurbishing/Repairing involves restoring returned or used goods to working condition, which may include repairs, cleaning, and minor alterations to ensure the product is functional as original. It extends the product lifecycle, thereby minimizing waste and lessening environmental impact. Predictive maintenance, supported by IoT and machine learning, identifies issues early and optimizes repair processes (Liu et al., 2023; Mosallanezhad et al., 2023). This practice is particularly effective in industries like electronics, automotive, and heavy machinery (Mejía-Moncayo et al., 2023).
- Recycling extracts value from end-of-life products by processing them into raw materials. It plays a critical role in industries such as plastics, metals, and electronics, reducing resource depletion and environmental harm (Bekrar et al., 2021; Villar et al., 2023). AI and blockchain technologies enhance transparency and efficiency in recycling systems, ensuring high-quality outputs and reducing contamination (Naseem et al., 2023; Fraga-Lamas et al., 2021).
- Remanufacturing restores products to "like-new" condition through extensive repairs or upgrades, commonly used in high-value industries like aerospace and defense. Digital twin technology helps simulate and optimize remanufacturing processes, reducing costs and maintaining quality standards (Tavana et al., 2023; Simonetto et al., 2022a).
- Disposal is the final step for products that cannot be reused, repaired, or recycled. Sustainable disposal methods, e.g., waste-to-energy technologies, minimize landfill waste while contributing to energy production (Franzè et al., 2023; Awan et al., 2021). Blockchain ensures transparency in the disposal process, verifying proper waste handling (Mo et al., 2022).

By integrating cutting-edge technologies, Industry 5.0 elevates reverse logistics, making it more efficient, sustainable, and aligned with circular economy principles (Lou et al., 2024; Sun et al., 2023). These advancements help industries achieve better environmental outcomes, seize new business opportunities, and enhance long-term profitability (Govindan, 2022).

8.3 Methodology

A systematic literature review (SLR) is conducted to identify, evaluate, and organize relevant studies on reverse logistics within Industry 5.0. An SLR enables a comprehensive, objective, and structured review, minimizing bias and providing a deeper understanding of the field (Kitchenham et al., 2009; Azarian et al., 2023). Denyer and Tranfield's (2009) five-stage model was followed: (1) formulating research questions, (2) locating studies, (3) selecting and evaluating studies, (4) analyzing and synthesizing data, and (5) reporting results, which help to better achieve the research objectives.

First, the research questions are defined as follows:

RQ1: What is the current research landscape of reverse logistics in Industry 5.0?

RQ2: Which reverse logistics practices can be enhanced by adopting Industry 5.0 technologies?

RQ3: How do these practices enhance sustainability across environmental, economic, and social dimensions?

The second step is to select targeted keywords to ensure a thorough and unbiased search. This process followed three phases to cover the primary areas of reverse logistics, Industry 5.0, and enabling technologies. Initially, keywords related to reverse logistics were identified, including terms such as "reverse logistics", "reverse supply chain", "product returns", "remanufacturing", and "recycling". These terms reflect essential reverse logistics processes, capturing activities centered on value recovery, returns, and disposal. Next, Industry 5.0 keywords are selected, which focus on concepts such as "Industry 5.0", "human-centricity", "resilience", and "sustainability". These terms align with Industry 5.0's emphasis on human-centered, sustainable, and resilient industrial systems, as defined by the European Commission (2021). Finally, relevant technologies associated with Industry 5.0 are identified, including "artificial intelligence", "big data analytics", "cobots", "digital twin", "internet of things", "cyber-physical system", "blockchain", and "edge computing". These technologies represent the key innovations driving Industry 5.0, enhancing efficiency, connectivity, and automation.

The third step is to locate the studies (data collection). The Scopus and Web of Science databases are used to collect relevant articles due to their extensive scholarly coverage and strong capabilities in citation and bibliometric analysis (Fahimnia et al., 2015; Pournader et al., 2021). A systematic search strategy was implemented by combining keywords across three main themes: reverse logistics, Industry 5.0, and enabling technologies. The Boolean operator "OR" was used to combine keywords within each theme, while the "AND" operator linked the themes to refine the search. The final search string was as follows:

("reverse logistics" OR "reverse supply chain" OR "product returns" OR "remanufacturing" OR "recycling") AND ("industry 5.0" OR "I5.0" OR "human-centricity" OR "sustainability" OR "resilience") AND ("artificial intelligence" OR "big data analytics" OR "digitalization" OR "cobots" OR "digital twin" OR "Internet of Things" OR "cyber physical system" OR "blockchain" OR "edge computing").

The fourth step is study selection and evaluation. In May 2024, an article search was conducted in the Scopus and Web of Science databases using the selected keywords in the title, abstract, and keywords sections. To ensure quality, only peer-reviewed journal articles in English were included, while conference proceedings, book chapters, preprints, and non-English publications were excluded. The subject areas were restricted to relevant fields, including Computer Science, Engineering, Business, Management, Decision Sciences, Energy, Economics, and Social Sciences. After applying these criteria, 113 articles were retrieved from Scopus and 89 from Web of Science. Removing duplicates resulted in a total of 132 articles. The filtering process began with a preliminary review of titles and abstracts to assess thematic relevance, leading to the exclusion of 85 articles that did not align with the study's focus on reverse logistics in Industry 5.0. In a second round, full-text reviews were conducted, further excluding 14 articles that lacked detailed discussion on the application of reverse logistics within Industry 5.0. Ultimately, 33 articles were selected for analysis, based on which bibliometric analysis and content analysis were conducted. Figure 8.1 illustrates the article search, selection, and evaluation strategy, providing a clear overview of the systematic process used to identify and analyze relevant studies for this research.

The bibliometric and content analysis sections offer in-depth analyses of the reviewed papers in the final step. The bibliometric analysis was carried out to examine the trend of publication, distribution of sources, co-authorship analysis, and keyword co-occurrence analysis using the selected sample. Subsequently, a content analysis is conducted to examine how the application of various advanced Industry 5.0 technologies can enhance reverse logistics practices and contribute to improved sustainability outcomes. It also brings out opportunities and challenges and identifies future research directions. Finally, the review is summarized, and the paper is concluded.

8.4 Bibliometric Analysis

Bibliometric analysis is a quantitative analysis widely applied in academic research to measure and analyze the volume, impact, and evolution of scholarly work in a particular field (Ellegaard & Wallin, 2015). In recent years, this method has become increasingly popular in business and management research, where it is used to map research influence, identify knowledge gaps, and gain insights into the structure of scientific communities (Zupic & Čater, 2015). To examine reverse logistics within the Industry 5.0 context, bibliometric analysis offers a comprehensive overview of the field's growth, its interdisciplinary appeal, and the technology-driven advancements reshaping the landscape.

An analysis of publication trends from 2020 to early 2024 reveals significant growth, with articles increasing from a modest 2 in 2020 to 17 by 2023. This spike underscores the heightened academic and industry interest in applying Industry 5.0's sustainability, resilience, and human-centric values to reverse logistics—a shift likely influenced by technological advancements and a stronger

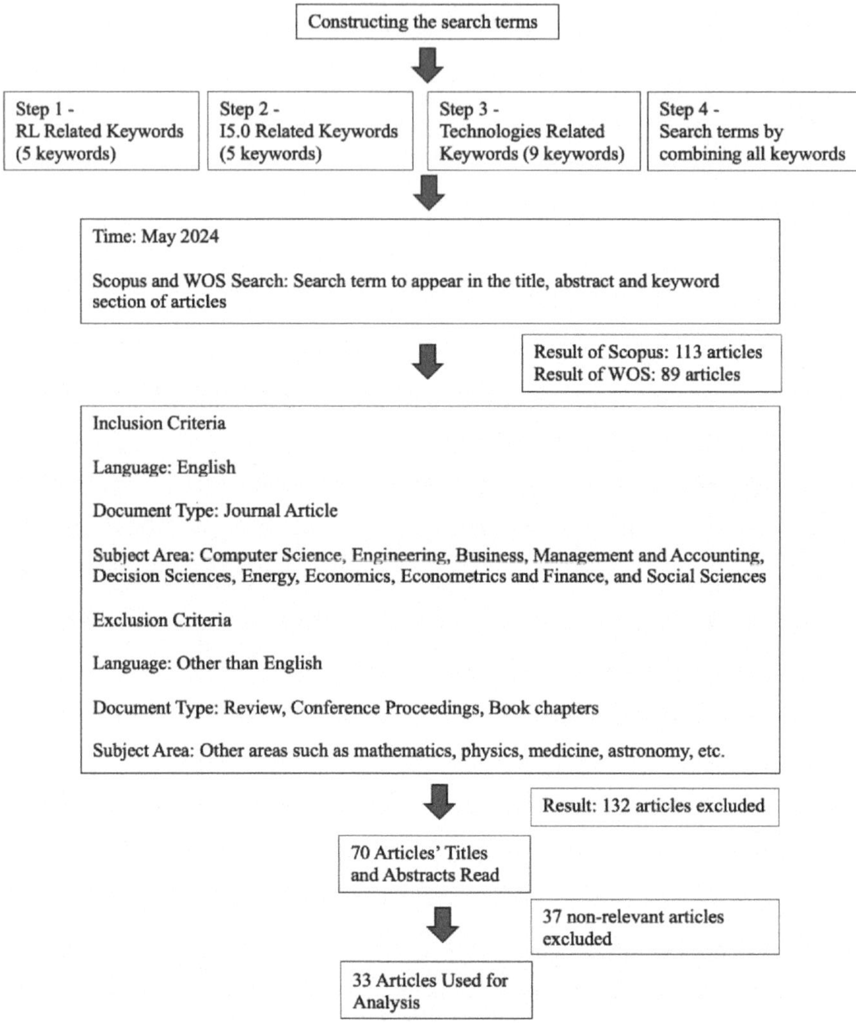

Fig. 8.1 Article search, selection, and evaluation strategy.

emphasis on circular economy practices. Although only one article was recorded by May 2024 due to early filtering, the increasing trend in past years suggests that research activity will likely continue to grow. Publisher analysis shows Elsevier and MDPI as major contributors, with Elsevier alone accounting for 34.5% of the publications, underscoring its prominent role in advancing the field. Other publishers, including Springer, Taylor & Francis, and Emerald, also contribute, reflecting a broad academic interest in reverse logistics within Industry 5.0.

The distribution of articles by journals highlights an interdisciplinary interest, with "Sustainability" leading the field, followed by "Computers and Industrial

Engineering" and "Business Strategy and the Environment". This distribution demonstrates that reverse logistics in Industry 5.0 spans topics like sustainability, engineering, business strategy, and AI. Highly cited works, including those by Awan et al. (2021) and Fraga-Lamas et al. (2021), showcase the field's focus on integrating technologies like IoT, AI, and blockchain to enhance sustainability. These works emphasize the value of Industry 5.0 technologies in improving reverse logistics processes, increasing transparency, and promoting sustainable practices. Collectively, the analysis shows a growing academic interest in how Industry 5.0's core principles: sustainability, resilience, and human-centricity, are advancing the field of reverse logistics, offering pathways to more sustainable and responsible supply chain models. Table 8.1 highlights the most influential works for reverse logistics in Industry 5.0.

Table 8.1 The most influential articles in the period of 2015-2024

Articles	*I5.0 Technologies*	*I5.0 Dimensions*	*Citations*
Awan et al. (2021)	IoT	Sustainability, Resilience and Human-centricity	238
Fraga-Lamas et al. (2021)	IoT and AI	Sustainability	120
Bekrar et al. (2021)	Blockchain	Sustainability and Resilience	89
Tozanlı et al. (2020)	IoT and Blockchain	None	72
Liu et al. (2023)	IoT, AI, Blockchain and Big Data	Sustainability	52
Govindan (2022)	Blockchain	Sustainability	51
Simonetto et al. (2022)	IoT and Big Data	Sustainability	49
Prajapati et al. (2022)	IoT and Blockchain	Sustainability	45
Alonso et al. (2021)	AI	Sustainability	36
Meier et al. (2023)	Blockchain	Sustainability	33

The keyword co-occurrence analysis in Figure 8.2 offers a comprehensive view of the central themes and interconnected research areas within the domains of reverse logistics and Industry 5.0. Using "all keywords" and "full counting" settings, the analysis mapped the relationships among 40 prominent keywords from previous studies, chosen from a total of 278 based on a minimum threshold of one co-occurrence. This network illustrates the core concepts, technologies, and sustainability goals shaping the current research landscape. The most frequently occurring keywords, i.e., "circular economy", "sustainable development", "Internet of Things", and "blockchain", highlight a strong research focus on sustainability and technological integration. For example, "circular economy" (Occurrence = 13, TSL = 197) and "sustainable development" (Occurrence = 11, TSL = 4) reflect a priority on achieving environmental and economic sustainability within Industry

5.0 frameworks, while the presence of keywords like "Internet of Things" and "blockchain" emphasize the essential role of digital technologies in facilitating these sustainable practices.

Table 8.2 Keyword co-occurrence analysis of reverse logistics in Industry 5.0

Keyword	*Occurrences*	*Total Link Strength*
circular economy	13	197
sustainable development	11	4
Internet of Things	8	131
blockchain	6	93
decision making	6	98
recycling	6	102
reverse logistics	6	45
supply chain management	6	67
Industry 4.0	4	69
closed-loop supply chain	3	53
Industry 5.0	3	66

These keywords are organized into nine distinct clusters, each representing specific research focuses that collectively capture the interdisciplinary nature of reverse logistics in Industry 5.0. Cluster 1 (blue) centers on the core dimensions of Industry 5.0, emphasizing human-centricity, resilience, and sustainability. Cluster 2 (purple) focuses on reverse logistics practices to enhance sustainability goals. Cluster 3 (red) highlights the role of artificial intelligence and waste management in advancing recycling and circular economy practices. Cluster 4 (green) discusses Industry 4.0 technologies like blockchain in closed-loop supply chains. Cluster 5 (yellow) includes optimization models using IoT and digital transformation. Cluster 6 (brown) examines decision-making processes enhanced by digital technologies. Cluster 7 (light blue) involves supply chain management and remanufacturing. Cluster 8 (orange) focuses on the digital circular economy's role in reducing carbon footprints. Cluster 9 (pink) relates to circular supply chains.

These clusters include dimensions of Industry 5.0 (such as human-centricity, resilience, and sustainability), reverse logistics practices that support sustainability, and key technologies like AI, IoT, and blockchain, which play critical roles in advancing recycling, closed-loop supply chains, and digital transformation. In addition, these clusters emphasize areas like supply chain management, decision-making enhanced by digital tools, remanufacturing, and the digital circular economy's impact on reducing carbon footprints. This result reveals how Industry 5.0 builds on the technological foundations established by Industry 4.0, while simultaneously aligning with sustainability and circular economy principles.

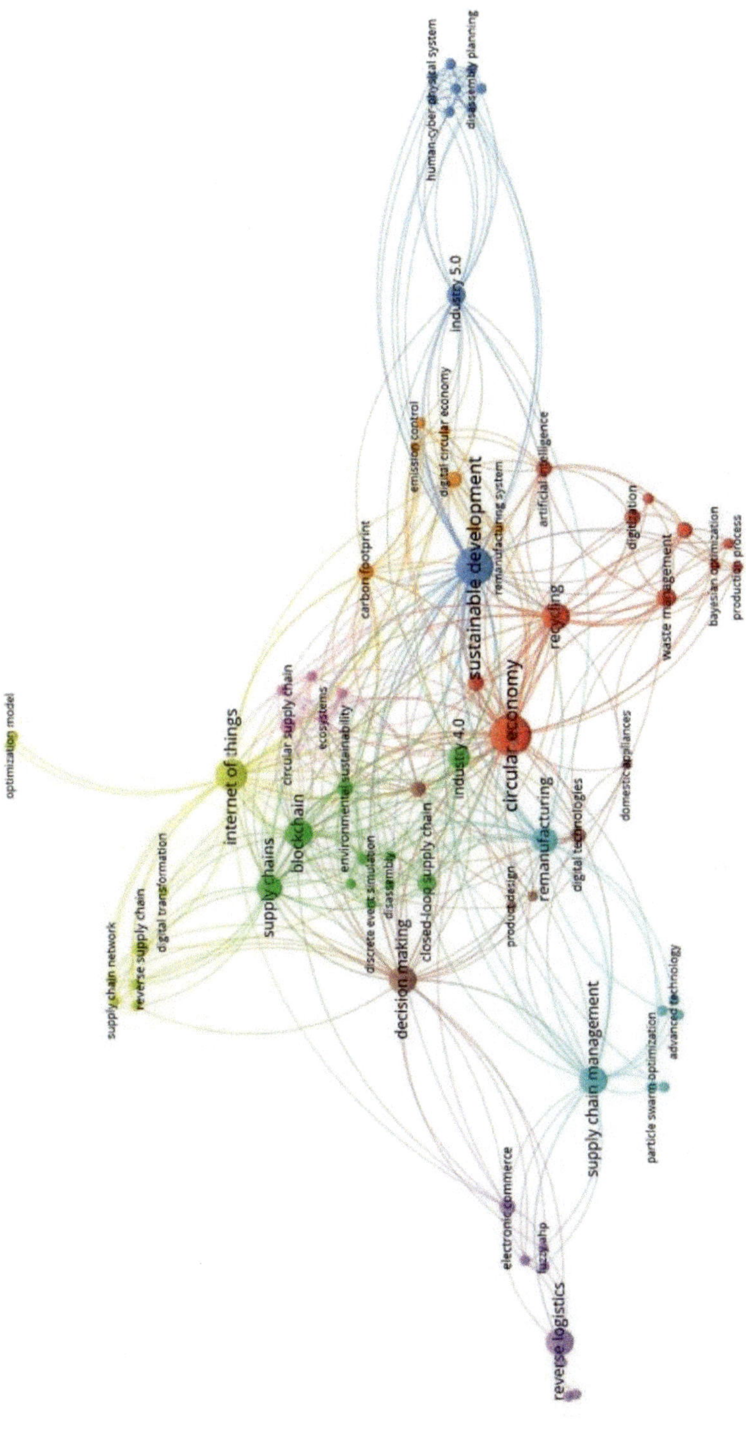

Fig. 8.2 Keyword co-occurrence analysis.

8.5 Content Analysis

This section provides an in-depth content analysis exploring the critical intersections between reverse logistics practices and the sustainability goals inherent in Industry 5.0. Through conducting a detailed content analysis of the selected articles to summarize the contributions of several related topical areas, we discuss how reverse logistics supports the three sustainability dimensions—environmental, social, and economic—and aligns with the human-centered and technology-integrated objectives of Industry 5.0. By focusing on reverse logistics practices, the analysis highlights the potential of Industry 5.0 technologies to enhance these practices, thereby creating a sustainable and resilient supply chain framework (Nahavandi, 2019).

The analysis also identifies six key reverse logistics practices: collection, reuse, repair, recycling, remanufacturing, and disposal, as foundational components of a circular economy in the Industry 5.0 landscape. These practices contribute not only to waste reduction but also to resource optimization, social sustainability, and economic efficiency, providing a basis for organizations to achieve long-term sustainability goals.

8.5.1 Reverse Logistics Practices within Industry 5.0

Industry 5.0 represents a shift towards integrating advanced technologies with a strong human-centered focus, emphasizing sustainability and resilience across supply chains. In this paradigm, reverse logistics practices play a vital role by ensuring that materials and products at the end of their life cycle are effectively managed, reused, and recycled (Dabo and Hosseinian-Far, 2023). A comprehensive content analysis of 33 studies reveals the emphasis on six primary reverse logistics practices, each with specific contributions to sustainability:

- *Collection:* Recognized as a critical process, efficient collection systems are crucial to minimizing landfill waste and retrieving valuable materials. Industry 5.0's technologies, such as IoT and AI, enable precise tracking and real-time monitoring of product returns, optimizing collection routes and reducing associated costs and emissions (Nahavandi, 2019). This efficiency not only minimizes environmental impact but also provides a structured approach to resource recovery, thereby enhancing both environmental and economic sustainability.
- *Reuse:* The reuse of materials and products extends their lifecycle and minimizes waste, an essential component of a sustainable supply chain. Content analysis indicates that 27% of studies discuss reuse, underscoring its role in reducing raw material dependency and fostering responsible consumption. With Industry 5.0 technologies like AI and IoT, companies can monitor product usage and determine optimal reuse opportunities, facilitating sustainable consumption patterns and contributing to the circular economy.
- *Refurbishing/Repair:* Surprisingly, refurbishing or repair processes are underrepresented in the literature, appearing in only 6% of studies. This gap

suggests an overlooked area within reverse logistics despite its potential for extending product lifespans and reducing environmental impact. Industry 5.0 technologies, such as robotics and AI-enabled diagnostics, can enhance repair processes, making them more accurate and cost-effective. This optimization can lead to significant environmental and economic benefits, positioning repair as a valuable component of a sustainable supply chain (Choudhary et al., 2020).

- *Recycling:* Recycling is the most extensively discussed practice, with 64% of studies emphasizing its importance in waste reduction and resource recovery. Recycling reduces reliance on virgin materials and minimizes carbon emissions, aligning with environmental sustainability goals. In Industry 5.0, AI-driven sorting technologies improve recycling accuracy and reduce contamination, enhancing the efficiency and sustainability of recycling processes (Govindan et al., 2015).
- *Remanufacturing:* As discussed in 39% of studies, remanufacturing restores products to near-original conditions, reducing the need for new material production and conserving energy. Industry 5.0's integration of robotics and 3D printing in remanufacturing processes allows for precision and efficiency, making remanufacturing a cost-effective and sustainable alternative to traditional production.
- *Disposal:* While disposal is considered a last resort, it remains a necessary component of reverse logistics. Content analysis shows that sustainable disposal practices, e.g., waste-to-energy processes, are becoming increasingly relevant. These methods not only reduce environmental impact but also contribute to economic sustainability by generating renewable energy from waste materials (Demirel & Kesidou, 2019).

Table 8.3 presents an overview of reverse logistics practices within the Industry 5.0 framework, highlighting their alignment with sustainability objectives and technological advancements. It categorizes key practices, emphasizing their roles in promoting circular economy principles. By leveraging Industry 5.0 technologies, these practices are enhanced to achieve greater efficiency, transparency, and sustainability. The table serves as a valuable reference for understanding how reverse logistics integrates with Industry 5.0 to create human-centered, resilient, and sustainable supply chains.

8.5.2 Impact of Industry 5.0 Technologies on Sustainable Reverse Logistics

Industry 5.0 introduces a range of advanced technologies that enhance the efficiency and sustainability of reverse logistics practices. Content analysis reveals several pivotal technologies in this context, with blockchain and IoT emerging as the most frequently discussed in the literature.

- *Blockchain:* With a focus on 48% of studies, blockchain is widely regarded for its potential to enhance transparency, traceability, and trust in reverse logistics.

Table 8.3 Reverse logistics practices in Industry 5.0

Authors	*Reverse logistics practices*					
	Collection	*Reuse*	*Repair*	*Recycling*	*Remanufacturing*	*Disposal*
(Lou et al. 2024)				✓		
(Sun, Yu & Solvang 2023)	✓			✓		✓
(Al Doghan & Sundram 2023)				✓		✓
(Tavana et al. 2023)	✓			✓		✓
(Jraisat et al. 2023)						
(Salikhov, Hayrutdinov & Muminov 2023)					✓	
(Meier, Gruchmann & Ivanov 2023)	✓	✓		✓		
(Hau et al. 2023)	✓	✓		✓		
(Ivanova & Shkrobot 2023)					✓	
(Eladly et al. 2023)				✓		
(Liu, Song & Liu 2023)				✓		
(Mejía-Moncayo, Kenné & Hof 2023)				✓	✓	
(Kazancoglu et al. 2023)						
(Franzè et al. 2023)						
(Wiegand & Wynn 2023)	✓	✓	✓	✓	✓	
(Mosallanezhad et al. 2023)	✓			✓		✓

Contd.

Contd.

Authors	*Reverse logistics practices*					
	Collection	*Reuse*	*Repair*	*Recycling*	*Remanufacturing*	*Disposal*
(Villar, Paladini & Buckley 2023)				✓		
(Naseem et al. 2023)						
(Ma, Qin & Hu 2022)	✓	✓		✓	✓	
(Chander et al. 2022)						
(Prajapati et al. 2022)	✓			✓	✓	✓
(Charnley et al. 2022)		✓		✓		
(Simonetto et al. 2022)	✓				✓	✓
(Mo et al. 2022)		✓	✓			✓
(Chaudhari et al. 2022)		✓		✓		✓
(Govindan 2022)					✓	
(Alonso et al. 2021)	✓				✓	
(Bekrar et al. 2021)	✓	✓		✓	✓	
(Fraga-Lamas, Lopes & Fernández-Caramés 2021)				✓		
(Awan, Sroufe & Shahbaz 2021)		✓		✓	✓	
(Bag et al. 2021)					✓	
(Kintscher et al. 2020)		✓		✓		
(Tozanlı, Kongar & Gupta 2020)				✓	✓	

By enabling secure tracking of materials across the supply chain, blockchain ensures accountability and fosters sustainable practices (Kouhizadeh, Saberi & Sarkis 2021).

- *IoT:* This technology is highlighted in 45% of the studies, which enables real-time data collection, tracking, and monitoring of products throughout their lifecycle. By optimizing collection, reuse, and remanufacturing processes, IoT supports resource efficiency and minimizes waste, aligning with Industry 5.0's sustainability goals.
- *AI:* Found in 30% of studies, AI improves recycling and repair processes by enabling precise sorting, predictive maintenance, and optimized resource management. AI-driven systems reduce human errors and increase operational efficiency, contributing to both environmental and economic sustainability (Bag et al., 2021).
- *Big data analytics:* Present in 18% of studies, big data analytics provides companies with valuable insights into product lifecycle management, waste reduction, and resource optimization. These insights help organizations make data-driven decisions, enhancing the sustainability of reverse logistics practices.

Conversely, technologies such as AM, AR/VR, Cybersecurity, and CPS are underrepresented in the literature. Despite their potential to revolutionize reverse logistics practices, these technologies are mentioned in only a few studies, indicating potential areas for further research.

8.5.3 Industry 5.0 Dimensions in Reverse Logistics Practices

Content analysis reveals that most studies focus on sustainability as a primary dimension of Industry 5.0, with 88% of papers emphasizing the environmental dimension. This trend highlights the significance of reverse logistics in promoting greener supply chains, reducing waste, and optimizing resource use. However, the analysis also identifies significant research gaps in resilience and human-centricity, two other core dimensions of Industry 5.0, within the context of reverse logistics.

- *Resilience:* Only a few studies discuss resilience in the context of reverse logistics, despite its relevance in adapting to disruptions. Reverse logistics practices, particularly collection and remanufacturing, could enhance supply chain agility and flexibility, making resilience an important area for future research. Industry 5.0 technologies, with real-time data and predictive analytics, could further strengthen resilience by allowing reverse logistics systems to adapt more quickly to unexpected disruptions, ensuring continuity even in volatile environments.
- *Human-Centricity:* Similarly, human-centricity is underexplored, with limited studies addressing how reverse logistics can enhance worker well-being and societal benefits. Industry 5.0's focus on collaboration between humans and intelligent systems presents opportunities to integrate more human-centered approaches into reverse logistics, e.g., improving working conditions and

supporting community involvement in sustainable practices. The potential for reverse logistics practices to foster community engagement, improve workplace ergonomics, and support a shared sense of responsibility toward sustainability represents a promising area for future research (Nahavandi, 2019).

- *Sustainability:* This dimension remains the most frequently addressed dimension in reverse logistics, aligning with Industry 5.0's goals of reducing environmental impact and promoting circular economy principles. By enabling processes such as recycling, remanufacturing, and resource recovery, reverse logistics contributes significantly to sustainable supply chains. Industry 5.0 technologies, like IoT and AI, allow for more efficient tracking, sorting, and processing of returned goods, thus optimizing resource recovery and minimizing waste. Within sustainability, reverse logistics can be explored through environmental, social, and economic lenses.

8.5.4 Environmental Sustainability Enabled by Reverse Logistics Practices

The first pillar is environmental sustainability regarding the decrease in the environmental impact from the industrial process. Industry 5.0 contributes by adding human-machine collaboration to reverse logistics, bringing more efficient and sustainable systems. These practices reduce waste, enhance resource efficiency, and cut carbon emissions on their own (Govindan et al., 2015).

- Collection makes it all important to retrieve such a product at the end of its life cycle, which would otherwise find itself in landfills. Industry 5.0 optimizes IoT technologies to make collection operations more efficient, most especially enhancing the real-time tracking of returned products and thereby reducing logistics costs and ensuing carbon emissions. Smart systems operating on AI can predict product returns, offering adequate plans for effective collection routes with lower energy use (Nahavandi, 2019).
- Reuse extends product life cycles by restoring used goods to working conditions, reducing the demand for new manufacturing. Industry 5.0 enhances reuse by utilizing AI-driven resource management tools that track materials' condition and assess whether they can be reused. This technology reduces waste and maximizes the lifecycle of products and materials (De Giovanni & Esposito Vinzi, 2012).
- Refurbishing/repair extends the product lifecycle and reduces the demand for new products. AI-enabled diagnostic tools and cobots under Industry 5.0 facilitate and speed up repair while making it more accurate. This also tends to reduce the environmental burden because of the reduced need for new components and, subsequently, basic materials (Nahavandi, 2019).
- Recycling is vital for reducing the demand for raw materials. Industry 5.0 has developed sorting technologies and advanced material recognition systems that improve the efficiency of recycling processes. Materials would have less

contamination, thus assuring a greater yield, and reducing energy and resource use in the recycling process (Demirel & Kesidou, 2019).

- Remanufacturing restores the product to a condition similar to the new with a drastic reduction of energy and material use compared to a new production process. Industry 5.0 makes remanufacturing viable through robotic automation and 3D printing technologies that make high-precision processes less wasteful and consume fewer resources while improving cost efficiency (Govindan et al., 2015).
- Disposal, although a last resort, is improved through waste-to-energy technologies in Industry 5.0, allowing companies to convert waste into renewable energy, hence reducing the ecological footprint of waste disposal and aligning with Industry 5.0 sustainability objectives (Demirel & Kesidou, 2011).

The technologies of Industry 5.0 enhance reverse logistics operations so that they reduce environmental impact and further circularity for world sustainable development.

8.5.5 Social Sustainability Enabled by Reverse Logistics Practices

Social sustainability in Industry 5.0 focuses on enhancing human well-being, improving the community's development, and handling ethical supply chain management. Reverse logistics play a major role in generating employment opportunities, improving working conditions, and enhancing ethical business practices (Bansal, 2005).

- Collection systems in Industry 5.0 will be further supplemented with AI and IoT, contributing to the improvement of working conditions through automation that is very burdensome to human capability. These technologies open up new employment opportunities in waste management and logistics, among others, particularly in data-driven operations that support job opportunities for both developed and emerging markets (Demirel & Kesidou, 2011).
- Reuse allows community involvement in the actions that will lead to sustainability. Industry 5.0 would provide the possibilities for returning the product to companies for re-use, with every incentive provided through AI-enabled rewarding systems and tracking through blockchain mechanisms that guarantee transparency and traceability. Such systems would contribute to creating a sustainable culture and responsible consumption among communities (Govindan et al., 2015).
- Refurbishing/Repair operations create specialized jobs focused on product restoration. Advanced diagnostics, combined with Industry 5.0 cobots, enable repair technicians to work smarter, with less physical demand and more precision. This technology supports local economies and fosters small repair businesses; these can deploy the tools of Industry 5.0 to compete in the global marketplace (Nahavandi, 2019).

- Recycling activities attributed to Industry 5.0 involve societies in the care of the environment. Automated plants using AI for recycling create new employment opportunities in processing waste at a higher level of technological sophistication and under safer conditions because the human element will not be exposed to harmful materials. Still, these technologies contribute to increasing public awareness and civic participation in sustainability causes (De Giovanni & Vinzi, 2012).
- Remanufacturing creates jobs, not only related to the dismantling and refurbishing of products but also to vocational training for advanced technologies. As Industry 5.0 merges robotics with AI, management and handling training will be provided to the workers for advanced and complex systems that will enhance competency development and social advancement (Govindan et al., 2015).
- Disposal in Industry 5.0 is handled responsibly using eco-friendly technologies such as waste-to-energy plants, transforming toxic waste into energy that can be used. This reduces the impact of waste on local communities and minimizes health risks, enhancing the overall quality of life (Bansal, 2005).

Overall, reverse logistics in Industry 5.0 contributes to social sustainability by creating jobs, community involvement, as well as encouraging ethical and responsible business operations.

8.5.6 Economic Sustainability Enabled by Reverse Logistics Practices

Besides, economic sustainability in Industry 5.0 refers to the long-term orientation toward profitability and competitiveness due to cost-saving technologies and resource optimization. Reverse logistics will have a crucial role in Industry 5.0 by saving operation costs, recovering value from discarded products, and unlocking new sources of revenue (De Giovanni & Vinzi, 2012).

- Collection processes in Industry 5.0 are driven through AI and IoT to optimize resource recovery and reduce operation costs. This system predicts the return of products, dynamically routing the collection vehicle, hence ensuring minimum use of fuel and labor while recovering valuable materials (Demirel & Kesidou, 2011).
- Reuse supports the economic sustainability of a firm by making it possible to create maximum value from its assets. Industry 5.0 will support reuse through analytics on data related to the condition and reusability of returned products so that the company can minimize resource consumption and lower production costs (Govindan et al., 2015).
- Refurbishing/Repair services offer considerable economic gains by extending product life cycles and cutting the cost of manufacturing new goods. Predictive Industry 5.0 maintenance powered by AI helps companies manage resources more effectively and reduce expensive repairs through early fault detection (Nahavandi, 2019).

- Recycling, apart from the extra revenue streams generated through the sale of recovered materials, reduces waste management costs. Industry 5.0 further improves profitability in recycling with its integration of AI-powered sorting and automation in recycling plants. This reduces labor costs and increases efficiency and therefore reduces raw material acquisition costs.
- Remanufacturing offers significant economic benefits by reclaiming value from used products. In the context of Industry 5.0, advancements in robotics and 3D printing are set to enhance this process. These technologies enable flexible remanufacturing, allowing businesses to renovate products with great precision and reduced costs. This approach not only boosts profitability but also minimizes environmental impact (Govindan et al., 2015).
- Disposal, though not the most liked option, would still be economically optimized in Industry 5.0 since it makes use of waste-to-energy technologies that could allow companies to create secondary income in return for waste-to-energy conversion. It will help to lessen the cost of waste disposal and hence add to profitability (Demirel & Kesidou, 2011).

In summary, reverse logistics practices in Industry 5.0 will substantially contribute to economic sustainability through cost reduction, revenue generation, and resource optimization in tandem with the goal of long-term profitability of businesses.

All these reverse logistics practices contribute to three-dimensional sustainability but also enable the promotion of research into higher stages of efficiency in operations and competitiveness of an Industry 5.0 enterprise. In fact, reverse logistics can be used not only to reduce waste and preserve resources but also to help the community and ensure long-term profitability. Therefore, reverse logistics practices are entirely included in the goals of sustainability of Industry 5.0. Table 8.4 outlines the three dimensions of sustainability, offering a structured framework to analyze the environmental, social, and economic impacts of reverse logistics practices within the context of Industry 5.0.

Table 8.4 Three dimensions of sustainability

Authors	*Sustainability dimensions*		
	Environmental	*Social*	*Economic*
Lou et al. (2024)	✓		
Sun et al. (2023)	✓		
Al Doghan and Sundram (2023)	✓	✓	✓
Tavana et al. (2023)	✓	✓	✓
Jraisat et al. (2023)	✓	✓	✓
Salikhov et al. (2023)	✓	✓	✓
Meier et al. (2023)	✓	✓	✓
Hau et al. (2023)	✓	✓	✓

Contd.

Authors	*Sustainability dimensions*		
	Environmental	*Social*	*Economic*
Ivanova and Shkrobot (2023)	✓		✓
Eladly et al. (2023)	✓		
Liu et al. (2023)	✓	✓	
Mejía-Moncayo et al. (2023)	✓	✓	✓
Kazancoglu et al. (2023)			
Franzè et al. (2023)	✓		✓
Wiegand and Wynn (2023)	✓	✓	✓
Mosallanezhad et al. (2023)	✓	✓	✓
Villar et al. (2023)	✓	✓	✓
Naseem et al. (2023)	✓	✓	✓
Ma et al. (2022)	✓	✓	✓
Chander et al. (2022)		✓	✓
Prajapati et al. (2022)	✓	✓	✓
Charnley et. al (2022)	✓	✓	✓
Simonetto et al. (2022)	✓	✓	✓
Mo et. al (2022)	✓	✓	✓
Chaudhari et al. (2022)	✓	✓	
Govindan (2022)			✓
Alonso et al. (2021)	✓		
Bekrar et al. (2021)	✓	✓	✓
Fraga-Lamas et al. (2021)	✓	✓	
Awan et al. (2021)	✓		✓
Bag et al. (2021)	✓		
Kintscher et al. (2020)			
Tozanlı et al. (2020)	✓		✓

8.6 Discussions

This section first explores key insights and challenges of sustainable reverse logistics in the Industry 5.0 era, followed by suggestions for future research.

There has been a growing focus on enhancing the sustainability of reverse logistics systems using Industry 5.0 technologies. Global efforts have been directed towards advancing theoretical development, technology transfer, and industrial applications. Based on the content analysis, Figure 8.3 provides a summary of the impacts of Industry 5.0 technologies on the economic, environmental, and social aspects of sustainable reverse logistics management. The findings of this study

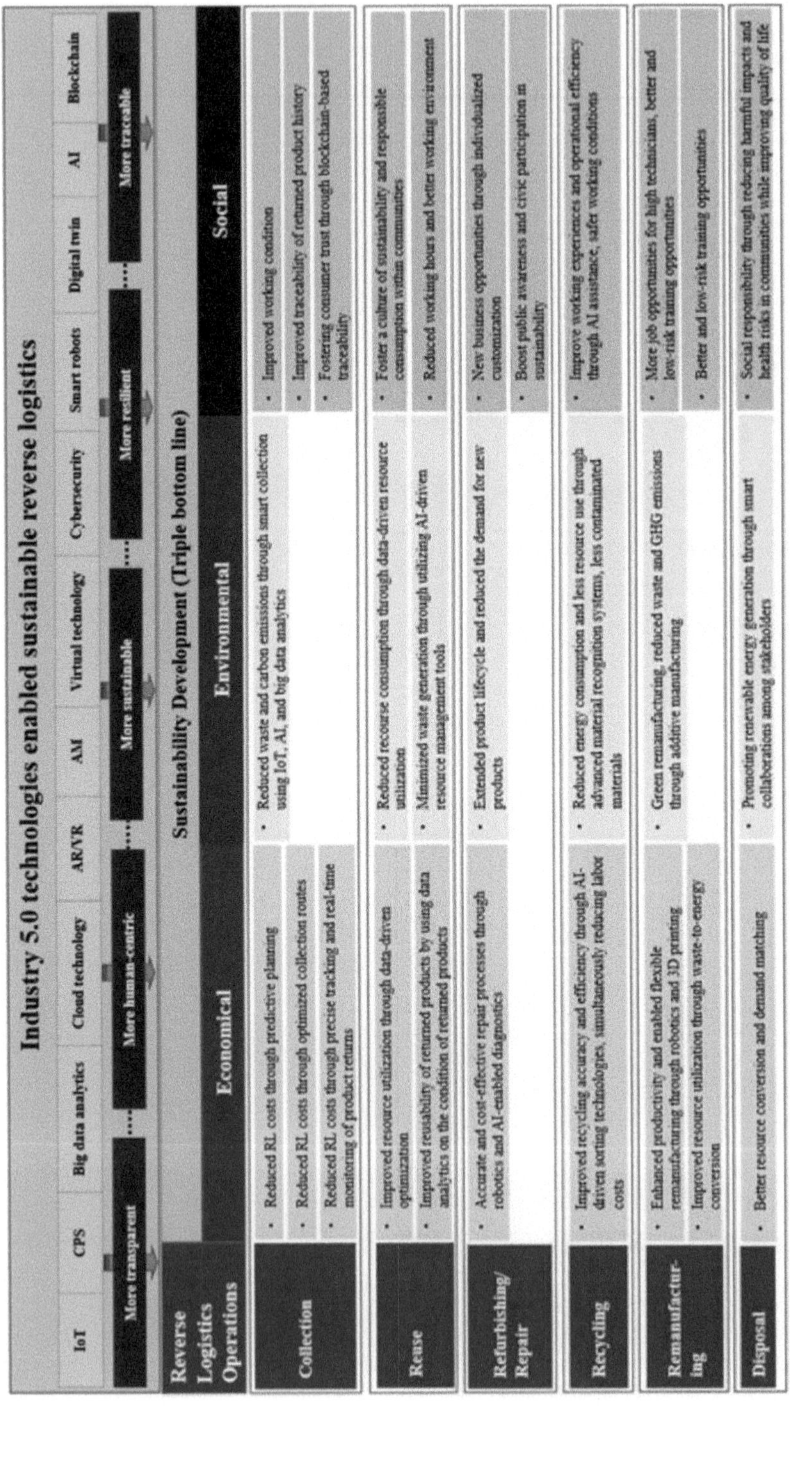

Fig. 8.3 Sustainable reverse logistics enabled by Industry 5.0 technologies.

highlight the transformative potential of integrating Industry 5.0 technologies into reverse logistics, emphasizing the shift toward human-centric, sustainable, resilient, and transparent supply chains. By incorporating enabling technologies such as AI, IoT, blockchain, cobots, digital twins, etc., reverse logistics processes can achieve unprecedented levels of efficiency and sustainability. However, the practical integration of these technologies into reverse logistics remains at an early stage, with notable gaps and challenges that require further exploration (Sun et al., 2022a).

One of the key insights is the shift from technology-oriented solutions, characteristic of Industry 4.0, to a balanced interplay of technology and human values in Industry 5.0. While technologies such as blockchain and IoT address traceability and transparency issues, cobots and digital twins emphasize the importance of human-machine collaboration. Despite these advancements, challenges persist in ensuring seamless integration, particularly in fostering human-centric approaches and achieving the sustainability goals central to Industry 5.0.

The discussion further underscores the role of reverse logistics in enabling circular economy models, which directly contribute to environmental sustainability by reducing waste and conserving resources. However, despite the significant amount of theoretical advancement, the limited number of empirical studies and practical implementations impedes a comprehensive understanding of the transformative potential of Industry 5.0 technologies in real-world reverse logistics operations.

In addition, a key tension emerges in aligning the economic benefits of reverse logistics with its environmental and social dimensions. This study has revealed that while technological solutions improve operational efficiency, the ethical and societal implications of reverse logistics practices remain underexplored. This gap represents an opportunity for future research to establish frameworks that harmonize technological innovation with ethical and social considerations.

Future research should focus more on empirical research, especially from primary data at the firm level, which is integrating Industry 5.0 technologies into the reverse logistics process. Case studies and field experiments could be conducted for deeper insight into real-world challenges and opportunities faced by the firm (Salikhov et al., 2023; Wiegand & Wynn, 2023). In addition, future research could also look in detail at how human-centered approaches might be more robustly integrated within reverse logistics, allowing human judgment to take a leading role in collaboration with advanced technologies, e.g., AI and cobots (Romero & Stahre, 2021). The other promising direction involves the investigation of the potentials of circular economy models to be opened by Industry 5.0 technologies, particularly in product return hotspots including, but not limited to, electronics, fashion, and automotive industries (Kazancoglu et al., 2023; Govindan, 2022) Future research could also involve how digital twin and blockchain technologies further extend the circle and how such technologies contribute to more efficient remanufacturing and recycling processes (Meier et al., 2023).

8.7 Conclusion and Future Research Suggestions

This chapter presents a systematic literature review focusing on the recent advancements and adoption of various Industry 5.0 technologies in sustainable reverse logistics management. First, a bibliometric analysis was conducted to identify publication trends, the most influential articles, and the most frequently used keywords. Then, a content analysis was carried out to examine the integration of Industry 5.0 technologies within reverse logistics to evaluate their potential opportunities to enhance sustainability across environmental, social, and economic dimensions. Six core reverse logistics practices, namely, collection, reuse, refurbishing/repair, recycling, remanufacturing, and disposal are discussed to investigate how Industry 5.0 technologies improve these practices to contribute to the goals of the circular economy. Finally, the paper summarizes current research developments and discusses the challenges, literature gaps, and future research opportunities.

This chapter investigates the incorporation of reverse logistics and Industry 5.0 technologies, coming to a place of prominence, therefore, to revolutionize conventional logistics processes toward sustainability, resilience, and human-centricity. A systematic literature review was conducted across key leading databases on what was critically assessed as directions given for the near future of reverse logistics regarding cutting-edge technologies like AI, IoT, blockchain, digital twins, and collaborative robots (cobots). The key findings underlined that these technologies, especially when applied to reverse logistics practices have considerable benefits in enhancing efficiency, transparency, and traceability, thus supporting economic and environmental sustainability. The challenges identified in the chapter were the limited number of human-centered applications and the obstacles to fully integrating Industry 5.0 technologies into reverse logistics. These findings show that there is an increasing interest in the optimization of reverse logistics by human-machine collaboration, while the potential to incorporate human creativity and decision-making in the logistics process remains underexplored.

This study has also found research gaps in the literature, which include a lack of empirical studies and limited presence of additive manufacturing in the context of reverse logistics. Whereas most of the papers discussed issues related to the technological and operational aspects of reverse logistics, a few were found to discuss how such practices align with the core values of Industry 5.0 regarding sustainability and human-centeredness. These gaps provide several opportunities for future research to be conducted on methodological development and case studies in order to find out how the technologies in Industry 5.0 will operate in real-world environments. Furthermore, future research should also seek to address the gaps related to human-machine collaboration, investigate the role of circular economy models, and further refine the enabling frameworks that assure the sustainability, resilience, and ethical awareness of reverse logistics practices.

Declaration of Generative AI and AI-assisted Technology in the Writing Process

During the preparation of the manuscript, the authors used ChatGPT and DeepSeek to polish the scientific writing. After using these tools, the authors reviewed and edited the content as needed and take full responsibility for the content of the paper.

Disclosure

This chapter is an improved version of the first author's diploma thesis entitled "*Reverse Logistics in Industry 5.0: Leveraging Advanced Technologies for Sustainable Practices*" submitted to the University of Rome, Tor Vergata.

Acknowledgment

Dr. Andrea Appolloni acknowledges the funding of this study supported by the European Union—NextGenerationEU in two projects: (1) the first project in the framework of the GRINS—Growing Resilient, Inclusive and Sustainable Project (GRINS PE00000018—CUP E83C22004690001) and (2) the second project in the framework of the ECS 0000024 Rome Technopole—CUP B83C22002820006, NRP Mission 4 Component 2 Investment 1.5.

References

Azarian, M., Yu, H., Shiferaw, A.T. and Stevik, T.K., 2023. Do We Perform Systematic Literature Review Right? A Scientific Mapping and Methodological Assessment. Logistics, 7(4), p.89.

Alonso, S.L.N., Forradellas, R.F.R., Morell, O.P. and Jorge-Vazquez, J., 2021, 'Digitalization, circular economy and environmental sustainability: The application of artificial intelligence in the efficient self-management of waste', Sustainability (Switzerland), 13(4), 2092.

Awan, U., Sroufe, R. and Shahbaz, M., 2021, 'Industry 4.0 and the circular economy: A literature review and recommendations for future research', Business Strategy and the Environment, 30(4), 2038–2060.

Backman, J., Valtanen, K., Laikari. Arto and Vallivaara, V., 2017, 'T330'.

Bag, S., Luthra, S., Mangla, S.K. and Kazancoglu, Y., 2021, 'Leveraging big data analytics capabilities in making reverse logistics decisions and improving remanufacturing performance', International Journal of Logistics Management, 32(3), 742–765.

Bakkar, M.N., 2023, 'CyberSecurity Essentials for Industry 5.0', pp. 49–65.

Bansal, P., 2005, 'Evolving sustainably: A longitudinal study of corporate sustainable development', Strategic Management Journal, 26(3), 197–218.

Bekrar, A., Cadi, A.A. El, Todosijevic, R. and Sarkis, J., 2021, 'Digitalizing the closing-of-the-loop for supply chains: A transportation and blockchain perspective', Sustainability (Switzerland), 13(5), 2895.

Chander, B., Pal, S., De, D. and Buyya, R., 2022, 'Artificial Intelligence-based Internet of Things for Industry 5.0', Internet of Things, pp. 3–45, Springer Science and Business Media Deutschland GmbH.

Charnley, F., Knecht, F., Muenkel, H., Pletosu, D., Rickard, V., Sambonet, C., Schneider, M. & Zhang, C., 2022, 'Can Digital Technologies Increase Consumer Acceptance of Circular Business Models? The Case of Second Hand Fashion', Sustainability (Switzerland), 14(8), 4589.

Chaudhari, R.S., Mahajan, S.K., Rane, S.B. and Agrawal, R., 2022, 'Modeling Barriers in Circular Economy Using TOPSIS: Perspective of Environmental Sustainability & Blockchain-IoT Technology', International Journal of Mathematical, Engineering and Management Sciences, 7(6), 820–843.

Chen, X., Eder, M.A. and Shihavuddin, A., 2020, 'A concept for human-cyber-physical systems of future wind turbines towards Industry 5.0'.

Cheng, J., Mahinder Singh, H.S., Zhang, Y.C. and Wang, S.Y., 2023, 'The impact of business intelligence, big data analytics capability, and green knowledge management on sustainability performance', Journal of Cleaner Production, 429, 139410.

Choudhary, T., Mishra, V., Goswami, A. and Sarangapani, J., 2020, 'A comprehensive survey on model compression and acceleration', Artificial Intelligence Review, 53(7), 5113–5155.

Dabo, A.-A. A. and Hosseinian-Far, A. 2023. An Integrated Methodology for Enhancing Reverse Logistics Flows and Networks in Industry 5.0. Logistics, 7, 97.

Demirel, P. and Kesidou, E., 2019, 'Sustainability-oriented capabilities for eco-innovation: Meeting the regulatory, technology, and market demands', Business Strategy and the Environment, 28(5), 847–857.

Denyer, D. and Tranfield, D., 2009, Producing a systematic review, Sage Publications Ltd.

Dev, N.K., Shankar, R. and Qaiser, F.H. 2020. Industry 4.0 and circular economy: Operational excellence for sustainable reverse supply chain performance. Resources, Conservation and Recycling, 153, 104583.

Doghan, M.A. Al and Sundram, V.P.K., 2023, 'AI-Enabled Reverse Logistics and Big Data for Enhanced Waste and Resource Management', Operational Research in Engineering Sciences: Theory and Applications, 6(2), 315–325.

Douissard, J., Hagen, M. E. and Morel, P. 2019. The da Vinci Surgical System. In: DOMENE, C. E., KIM, K.C., VILALLONGA PUY, R. & VOLPE, P. (eds.) Bariatric Robotic Surgery: A Comprehensive Guide. Cham: Springer International Publishing.

Eladly, A.M., Abed, A.M., Aly, M.H. and Salama, W.M., 2023, 'Enhancing circular economy via detecting and recycling 2D nested sheet waste using Bayesian optimization technique based-smart digital twin', Results in Engineering, 20, 101544.

Ellegaard, O. and Wallin, J.A., 2015, 'The bibliometric analysis of scholarly production: How great is the impact?', Scientometrics, 105(3), 1809–1831.

Fraga-Lamas, P., Lopes, S.I. and Fernández-Caramés, T.M., 2021, 'Green iot and edge AI as key technological enablers for a sustainable digital transition towards a smart circular economy: An industry 5.0 use case', Sensors, 21(17), 5745.

Franzè, C., Pesce, D., Kalverkamp, M. and Pehlken, A., 2023, "Scale without mass": A decision-making tool for scaling remanufacturing practices in the white goods industry', Journal of Cleaner Production, 417, 138078.

Gamage, H.T.M., Weerasinghe, H.D. and Dias, N.G.J., 2020, 'A Survey on Blockchain Technology Concepts, Applications, and Issues', SN Computer Science, 1(2), 114.

Giovanni, P. De and Esposito Vinzi, V., 2012, 'Covariance versus component-based estimations of performance in green supply chain management', International Journal of Production Economics, 135(2), 907–916.

Govindan, K., 2022, 'Tunneling the barriers of blockchain technology in remanufacturing for achieving sustainable development goals: A circular manufacturing perspective', Business Strategy and the Environment, 31(8), 3769–3785.

Govindan, K., Soleimani, H. and Kannan, D., 2015, 'Reverse logistics and closed-loop supply chain: A comprehensive review to explore the future', European Journal of Operational Research, 240(3), 603–626.

Hau, B.M., You, S.S., Cho, G.S., Yeon, J.H. and Kim, H.S., 2023, 'Circular Supply Chain Management with Blockchain Integration', Logforum, 19(4), 515–533.

Ivanova, T. and Shkrobot, M., 2023, 'Digitalization in the Reverse Supply Chain: A Bibliometric Analysis', Logforum, 19(4), 683–702.

Ivanov, D. 2023. The Industry 5.0 framework: viability-based integration of the resilience, sustainability, and human-centricity perspectives. International Journal of Production Research, 61, 1683-1695.

Jraisat, L., Jreissat, M., Upadhyay, A. and Kumar, A., 2023, 'Blockchain Technology: The Role of Integrated Reverse Supply Chain Networks in Sustainability', Supply Chain Forum, 24(1), 17–30.

Kazancoglu, Y., Ozbiltekin-Pala, M., Sezer, M.D., Luthra, S. and Kumar, A., 2023, 'Resilient reverse logistics with blockchain technology in sustainable food supply chain management during COVID-19', Business Strategy and the Environment, 32(4), 2327–2340.

Kintscher, L., Lawrenz, S., Poschmann, H. and Sharma, P., 2020, 'Recycling 4.0-digitalization as a key for the advanced circular economy', Journal of Communications, 15(9), 652–660.

Kitchenham, B., Pearl Brereton, O., Budgen, D., Turner, M., Bailey, J. and Linkman, S., 2009, 'Systematic literature reviews in software engineering – A systematic literature review', Information and Software Technology, 51(1), 7–15.

Kouhizadeh, M., Saberi, S. and Sarkis, J., 2021, 'Blockchain technology and the sustainable supply chain: Theoretically exploring adoption barriers', International Journal of Production Economics, 231, 107831.

Krstić, M., Agnusdei, G.P., Miglietta, P.P., Tadić, S. and Roso, V. 2022. Applicability of industry 4.0 technologies in the reverse logistics: a circular economy approach based on comprehensive distance based ranking (COBRA) method. Sustainability, 14, 5632.

Liu, L., Song, W. and Liu, Y., 2023, 'Leveraging digital capabilities toward a circular economy: Reinforcing sustainable supply chain management with Industry 4.0 technologies', Computers and Industrial Engineering, 178, 109113.

Lou, S., Zhang, Y., Tan, R. and Lv, C., 2024, 'A human-cyber-physical system enabled sequential disassembly planning approach for a human-robot collaboration cell in Industry 5.0', Robotics and Computer-Integrated Manufacturing, 87, 102706.

Lu, Q., Ren, Y., Jin, H., Meng, L., Li, L., Zhang, C. and Sutherland, J.W., 2020, 'A hybrid metaheuristic algorithm for a profit-oriented and energy-efficient disassembly sequencing problem', Robotics and Computer-Integrated Manufacturing, 61, 101828.

Luciano, C., Banerjee, P., Florea, L. and Dawe, G., 2005, Design of the ImmersiveTouch TM : a High-Performance Haptic Augmented Virtual Reality System.

Ma, D., Qin, H. and Hu, J., 2022, 'Achieving triple sustainability in closed-loop supply chain: The optimal combination of online platform sales format and blockchain-enabled recycling', Computers and Industrial Engineering, 174, 108763.

Meier, O., Gruchmann, T. and Ivanov, D., 2023, 'Circular supply chain management with blockchain technology: A dynamic capabilities view', Transportation Research Part E: Logistics and Transportation Review, 176, 103177.

Mejía-Moncayo, C., Kenné, J.P. and Hof, L.A., 2023, 'On the development of a smart architecture for a sustainable manufacturing-remanufacturing system: A literature review approach', Computers and Industrial Engineering, 180, 109282.

Mo, D.Y., Ma, C.Y.T., Ho, D.C.K. and Wang, Y., 2022, 'Design of a Reverse Logistics System with Internet of Things for Service Parts Management', Sustainability (Switzerland), 14(19), 12013.

Mosallanezhad, B., Gholian-Jouybari, F., Cárdenas-Barrón, L.E. and Hajiaghaei-Keshteli, M., 2023, 'The IoT-enabled sustainable reverse supply chain for COVID-19 Pandemic Wastes (CPW)', Engineering Applications of Artificial Intelligence, 120, 105903.

Nahavandi, S., 2019, 'Industry 5.0-a human-centric solution', Sustainability (Switzerland), 11(16), 4371.

Naseem, M.H., Yang, J., Zhang, T. and Alam, W., 2023, 'Utilizing Fuzzy AHP in the Evaluation of Barriers to Blockchain Implementation in Reverse Logistics', Sustainability (Switzerland), 15(10), 7961.

Prajapati, D., Jauhar, S.K., Gunasekaran, A., Kamble, S.S. and Pratap, S., 2022, 'Blockchain and IoT embedded sustainable virtual closed-loop supply chain in E-commerce towards the circular economy', Computers and Industrial Engineering, 172, 108530.

Rajput, S. and Singh, S.P. 2022. Industry 4.0 model for integrated circular economy-reverse logistics network. International Journal of Logistics Research and Applications, 25, 837-877.

Rani, S., Jining, D., Shoukat, K., Shoukat, M.U. and Nawaz, S.A., 2024, 'A Human–Machine Interaction Mechanism: Additive Manufacturing for Industry 5.0—Design and Management', Sustainability, 16(10), 4158.

Romero, D. and Stahre, J., 2021, 'Towards The Resilient Operator 5.0: The Future of Work in Smart Resilient Manufacturing Systems', Procedia CIRP, 104, 1089–1094.

Roy Chowdhury, R., Aneja, S., Aneja, N. and Abas, P.E., 2021, 'Packet-level and IEEE 802.11 MAC frame-level network traffic traces data of the D-Link IoT devices', Data in Brief, 37, 107208.

Salikhov, J., Hayrutdinov, S. and Muminov, T.K., 2023, 'Blockchain-Enabled Sustainable Supply Chain under Information Sharing and Recovery Quality Efforts', Sustainability (Switzerland), 15(5), 3929.

Sharma, M., Tomar, A. and Hazra, A., 2024, 'Edge Computing for Industry 5.0: Fundamental, Applications, and Research Challenges', IEEE Internet of Things Journal, 11(11), 19070–19093.

Simonetto, M., Sgarbossa, F., Battini, D. and Govindan, K., 2022, Closed loop supply chains 4.0: From risks to benefits through advanced technologies. A literature review and research agenda, International Journal of Production Economics, 253, 108582.

Siriwardhana, Y., Porambage, P., Liyanage, M. and Ylianttila, M., 2021, AI and 6G Security: Opportunities and Challenges, 2021 Joint European Conference on Networks and Communications & 6G Summit (EuCNC/6G Summit), 616–621, IEEE.

Sun, X., Yu, H. and Solvang, W.D., 2023, A Digital Reverse Logistics Twin for Improving Sustainability in Industry 5.0, IFIP Advances in Information and Communication Technology, vol. 690 AICT, 273–286, Springer Science and Business Media Deutschland GmbH.

Sun, X., Yu, H. and Solvang, W. D. System integration for smart reverse logistics management. 2022 IEEE/SICE International Symposium on System Integration (SII), 2022a. IEEE, 821–826.

Sun, X., Yu, H. and Solvang, W. D. 2022b. Towards the smart and sustainable transformation of Reverse Logistics 4.0: A conceptualization and research agenda. Environmental Science and Pollution Research, 29, 69275–69293.

Tavana, M., Khalili Nasr, A., Santos-Arteaga, F.J., Saberi, E. and Mina, H., 2023, 'An optimization model with a lagrangian relaxation algorithm for artificial internet of things-enabled sustainable circular supply chain networks', Annals of Operations Research, 342(1), 767-802.

Tozanlı, Ö., Kongar, E. and Gupta, S.M., 2020, 'Trade-in-to-upgrade as a marketing strategy in disassembly-to-order systems at the edge of blockchain technology', International Journal of Production Research, 58(23), 7183–7200.

Tromp, J., Le, D.-N. and Le, C., 2020, 'TOWARDS DIGITAL TWINS AND ROBOTICS', Emerging Extended Reality Technologies For Industry 4.0, pp. 97–99, Wiley.

Villar, A., Paladini, S. and Buckley, O., 2023, Towards Supply Chain 5.0: Redesigning Supply Chains as Resilient, Sustainable, and Human-Centric Systems in a Post-pandemic World, Operations Research Forum, 4(3).

Wang, L., Gao, R., Váncza, J., Krüger, J., Wang, X.V., Makris, S. and Chryssolouris, G., 2019, 'Symbiotic human-robot collaborative assembly', CIRP Annals, 68(2), 701–726.

Wang, L. and Ji, W., 2018, 'Cloud Enabled CPS and Big Data in Manufacturing', pp. 265–292.

Wiegand, T. and Wynn, M., 2023, 'Sustainability, the Circular Economy and Digitalisation in the German Textile and Clothing Industry', Sustainability (Switzerland), 15(11), 9111.

Zupic, I. and Čater, T., 2015, 'Bibliometric Methods in Management and Organization', Organizational Research Methods, 18(3), 429–472.

Index

A

AMPL 4, 5, 7, 9, 11, 21
AdaBoost 66, 67
Additive manufacturing (AM) 189
AI-driven 112, 128, 131, 141, 144, 153, 155, 158, 161, 168, 190, 199, 2020, 203
AI-driven insights 170
Algorithms 61, 62, 65-67
Analytical models 32,35
AnyLogistix 40, 41, 43-45, 49-52
Applications 31, 60-62, 65, 83, 87
Artificial intelligence (AI) 189
Artificial Neural Networks (ANNs) 82
Augmented Reality (AR) 117
Automated Guided Vehicle (AGV) 134
Automation 116, 163, 166
Autonomous delivery 115
Autonomous vehicles 115, 163

B

Bibliometric analysis 193
Big data analytics 113, 189, 202
Blockchain 103, 11, 113, 152, 160, 175, 179, 199

C

Capacitated network design 2, 5-14, 19, 27, 44, 50
Capacity constraint 7, 9, 13-15, 19-21, 23, 44, 46, 47, 50
Case studies 209
Center-of-gravity model 3
ChatGPT 15, 17, 18, 20, 21, 23, 25
Circular Economy (CE) 149, 151, 153, 155, 159, 162, 168, 172, 176, 180
Circular economy principles 149, 151, 196, 199
Circular packaging 157
Circular supply chain 158, 160, 161
Circularity 152-155, 157, 175, 178, 181
Closed-loop 149, 151, 160, 168-172
Closed-loop supply chain 160, 196
Cloud computing 34, 114, 115, 189
Collaborative logistics 119, 120
Collaborative robots (cobots) 159, 168, 173, 189
Collection 33, 38, 60, 69, 89, 202, 203, 205
Comments 17-21, 23-25
Compliance issues 140
Concepts 31, 99, 104, 155, 159, 162, 192, 195
Content analysis 187, 188, 193, 198, 199
Continuous improvement 163, 170, 172
Convolutional Neural Networks (CNNs) 84, 85
Cost barriers 136
Culture of innovation 177, 180, 181
Customer demand 1, 6, 13, 40, 42-44
Customer education 170
Customer engagement 168, 169, 171, 172
Cyber-physical systems (CPS) 138, 188
Cybersecurity 2, 140, 141, 190, 202

D

Data security 129, 135, 139, 140, 189
Data visualization 51
Data-driven 36, 57, 58, 60, 6, 159, 202
Decision-making 3, 5, 6, 19, 27, 35, 37, 38, 52, 60, 112, 135, 159
Demand forecasting 58, 60, 84, 88, 99, 111, 153
Digital logistics twins 2, 40, 52, 53
Digital twin 31-39, 41, 192, 209
Digital twins 32-38, 52, 131, 134, 187, 190, 209, 210
Digitalization 2, 36, 99, 192
Disposal 1, 158, 161, 175, 187, 190, 191
Distance matrix 8, 23, 52
Distribution 1-10, 27, 31, 41, 52, 96, 106, 193
Distribution network 1-7, 12, 13, 18, 19, 27, 40 41
Drones 115

E

Education 3, 16, 26, 27, 53, 103, 106, 170, 174
Employee empowerment 162
End-of-life 151, 155, 159-161, 191
Ensemble learning models 57, 58
Environmental issues 103
Extended reality (XR) 190

F

Feedforward Neural Networks (FNNs) 83
Financial considerations 175, 176, 178
Financial support and incentives 178
Freight transport 107
Future research suggestions 188, 210

G

Gaussian mixture model (GMM) 61, 62
GBDT (Gradient-Boosted Decision Tree) 67, 76
Generative Adversarial Networks (GANs) 84
Generative AI 1, 2, 15, 53, 78, 99
Gradient Boosting Regression Trees (GBRT) 59, 70
Gravity location model 2-5, 8, 27, 40, 52
Green logistics 118
Greenfield analysis (GFA) 40

H

High-density areas 76
Historic demand 42, 45
Human-centric 32, 52, 121, 138, 143, 145, 146, 151, 163
Human-centric repair 168, 172
Human-centricity 2, 99, 153-155, 169, 172, 186, 195
Human-machine collaboration 130, 158, 159, 162, 203, 209, 210
Human-Robot Collaboration (HRC) 129, 189

I

IIoT 143, 144
Industry 4.0 32, 128-130, 132, 150, 180
Industry 5.0 2
Industry 5.0 dimensions 202
Industry 5.0 technologies 99, 150, 151, 156, 168, 187, 192, 195, 203, 207, 210
Industry collaboration 170, 172
Information and communication technology (ICT) 36
Integrated system 32, 33
Integration 2, 16, 31-34, 36, 37, 40, 144, 163, 210
Internet of Things (IoT) 60, 110, 188

L

Large Language Models (LLMs) 2, 15
Layout and planning 133
Logistics network design 1, 2, 3, 5, 16, 26-28, 31
Logistics planning 36, 57, 99
Long Short-Term Memory (LSTM) Models 87

M

Maintenance1 108, 110, 119, 129, 133, 134, 142, 165, 188, 189
Manufacturing 2, 32, 35, 38, 41, 170
Mathematical models 18, 84
Methods 1, 2, 31, 33, 40, 58, 60, 61, 65, 66, 76, 82
Mixed Reality (MR) 128
Model framework 90
Models 1, 2, 15, 18, 23, 26, 82, 84, 85, 87, 94, 185, 195, 209

N

Natural language processing (NLP) 15
Network design 1-10, 44-46, 50, 52
Network Optimization (NO) 43, 44
Numerical example 2, 4, 41

O

On-site 132-134
Optimization 2-6, 14, 24, 27, 35, 38, 43, 52, 82, 96, 131, 146, 199, 202

P

Partial Dependence Plots (PDP) 76
Penalty 13, 47
Periodic demand 45
P-median model 18, 19
Practical application 61, 63, 66, 86, 92, 118, 136
Privacy challenges 139
Prompt 17-21, 23, 24, 32, 141
Pros and cons 61, 62, 66, 67, 83, 84

R

Random Forest (RF) Model 67
Real-time communication 114, 188
Real-time data 32, 35, 38, 84, 99, 106, 110, 115, 131, 134, 136, 139, 143
Recurrent Neural Networks (RNNs) 83, 87
Recycling 1, 37, 149, 151, 152, 161, 169-171, 175, 179, 179
Refurbishing 151, 153, 155, 170, 175, 190, 191, 198, 203, 205, 210
Regulatory 130, 135, 140, 141
Remanufacturing 37, 151, 153, 155, 187, 190-192, 199, 200
Remote operations 134
Repair 133, 158, 168-170, 172, 187, 190, 191, 198, 203
Research agenda 186
Resilience 2, 37, 38, 59, 60, 110, 121, 129, 154, 155, 163, 172, 179

Resource consumption 150, 151, 155, 173, 205
Response summary 17-21, 23-25
Responsiveness 1, 32, 36, 52, 99, 110, 112, 139, 165
Reusable solutions 157
Reuse 149, 152, 153, 157, 158, 161, 179, 186, 187, 190, 191, 198, 201, 203, 204, 205, 210
Reverse Logistics 158, 159, 186, 188, 198, 199, 203-205
Reverse logistics framework 188
Reverse logistics practices 200-202
RF (Random Forest) 66
Robotics 116

S

Safety operations 133
Service logistics 157
Simulation 32, 33, 35, 36, 38, 40, 52, 117, 130, 131, 133, 134, 138, 142
Single-source 2, 8, 9, 10, 14, 15, 23, 27, 53
Single-use packaging 156, 157, 170, 171
Skills development 173-178, 180, 181
Smart logistics 109, 110
Smart robots 32, 189
Smart sensors 99
Smart transportation 57-59, 61, 65, 67, 69, 71, 73, 75, 77, 78
Social partnerships 119
Social value 103, 104
Solar energy 118
Strategic planning 31, 165
Supply chain management 31, 33, 35-38, 40, 88, 129, 131, 145, 146, 155, 157, 176, 189, 196
Supply chain partners 154, 160
Supply chain resilience 37, 175-177, 179
Supply chains 2, 37, 38, 60, 72, 73, 104, 105, 111, 112, 119, 151, 203
Sustainability 2, 24, 32, 36, 37, 38, 52, 57, 59, 60, 99, 104-114, 118
Sustainable Development Goals (SDGs) 105
Sustainable Logistics 103, 107, 113, 115, 118, 119, 149, 151, 153, 179, 181
Sustainable packaging solutions 155, 157, 158
Sustainable practices 32, 36, 104, 114, 120, 128, 129, 158, 160, 161, 178, 181, 187
Sustainable reverse logistics 188, 199, 207, 208, 210
Sustainable secondary packaging 157
Sustainable Transport 103-106, 109, 120
Systematic literature review 186, 187, 191, 210

T

Theoretical framework 92
Training challenges 138
Transformation 32, 36, 128, 131, 145, 196

U

Uncapacitated network design 2, 13, 14, 15, 27, 50, 52
Unsupervised learning models 57, 58
Urban logistics 107, 111
User acceptance 138

V

Value creation 37
Virtual Reality (VR) 117

W

Warehousing 128, 129, 131, 134, 135, 141
Wearable technology 117
Worker training 133

X

XGBoost 66
XR Implementation 135, 136

Y

YOLOv8 model 94, 95, 98